Too Much Luck

TOO MUCH LUCK

THE MINING BOOM AND AUSTRALIA'S FUTURE

Paul Cleary

Black Inc.

Published by Black Inc.,
an imprint of Schwartz Media Pty Ltd
37–39 Langridge Street
Collingwood VIC 3066 Australia
email: enquiries@blackincbooks.com
http://www.blackincbooks.com

The National Library of Australia Cataloguing-in-Publication entry:

Cleary, Paul, 1964-

Too much luck : the mining boom and Australia's future / Paul Cleary.

ISBN: 9781863955379 (pbk.)

Natural resources--Australia. Australia--Economic conditions.
Australia--Economic conditions--Forecasting. Australia--
Social conditions. Australia--Social conditions--Forecasting.

330.994

Contents

Preface

On 361 mining complexes all over Australia, men, women and mammoth machines are operating around the clock on a scale never before seen in this country's history. Australia now produces more than one billion tonnes of minerals each year—and that's only counting the finished product. This is five times more than we were producing at the end of the last boom thirty years ago. It is enough to fill 3,000 of the biggest bulk carriers that ply the world's oceans, and the prices obtained from those shipments have risen on average by more than three times since the current boom got underway in 2003. While the impact of this boom is changing the face of our continent in ways that will be irreversible, our politicians lack the courage and capacity to put in place policies to ensure that we, and our descendants, reap a lasting legacy from this once-in-a-century opportunity.[1]

This current boom, the fifth in our history, may have only just begun. The demand for minerals to feed Asia's industrial and urban transformation is forecast by government experts to last decades. New industries like liquefied natural gas (LNG) have signed contracts to quadruple exports over the coming ten years and will soon rival coal and iron ore in export earnings. Despite this phenomenal change in our fortunes, our governments still have vast blind spots when it comes to managing the industry and the tax revenue it generates. Our leaders spend the income from our resource wealth like there's no tomorrow, even though this money is derived by running down our natural endowments. They are spending our capital. If we were the clever country we claim to be, we'd turn a share of this revenue into financial assets designed to last forever. Instead, our politicians behave like those of developing countries, spending indiscriminately and allowing mining companies to exert undue influence. So 'orestruck' are many of our state and federal leaders that they lack the will to tax and regulate this industry effectively.

To avoid the pitfalls experienced by other resource-rich nations over past decades, Australia needs to make changes in three areas: saving, taxation and regulation. We should put a substantial share of resource revenue into a fund that can be drawn on when boom turns to bust—as it always does—and to compensate future generations, who will have to get by with considerably less in the way of mineral wealth. Overseas experience in places like Norway, Chile and even tiny East Timor shows us how to do this, and we can improve on these models by 'pollie-proofing' our funds. These examples are there for us to learn from—but so far our leaders have shown little interest.

The global financial crisis starkly revealed weaknesses in how we manage our resource income. In the three years before the crisis, the federal government's coffers swelled by $334 billion in additional revenue, as low interest rates and the mining boom underpinned strong economic growth.[2]

Nearly all of it was spent on tax cuts and middle-class welfare by John Howard—pissed up against the wall, if the truth be told. This spending binge forced the Reserve Bank to jack up interest rates by 3 percentage points to 7.25 per cent. When the GFC arrived, the new Labor government ran up debt of $106 billion as it sought to stave off recession. And when natural disasters hit Queensland in early 2011, the government had to introduce a new levy to pay for the reconstruction because it didn't have a fund to draw on. By comparison, Chile, another resource-rich nation, had foreign-currency wealth funds that it had built up during its boom years. It used part of these savings to pay for an even bigger stimulus package and got through the GFC and a devastating earthquake without racking up a single peso of debt.

The really scary thing is that after this appalling episode we still don't have policies in place to prevent it from happening all over again. Without policies to save windfall and to compensate future generations, our state and federal treasurers will continue this feast-to-famine cycle. They will have put aside nothing to deal with unforseen events like natural disasters, the effects of climate change or war.

Without stronger and more effective government control, Australia will continue travelling at breakneck speed towards the bottom of the quarry, a journey that will wreak havoc on the non-resource sector and potentially leave many people far worse off. As the resource boom accelerates, it will keep the dollar sky-high and force up the cost of doing business for everyone else. Industries such as tourism and education—industries that, unlike mining, involve many jobs—will fade away as our currency soars even higher, propelled by speculative investment as the Australian dollar becomes Asia's new hard currency. We will be left high and dry if commodity prices suddenly collapse, as they did during the GFC in 2008, or worse, when the resources start to run out.

Our state and federal politicians have become so bedazzled by the prospect of even greater mineral riches that they are eagerly encouraging a resources rush while neglecting long-term ecological and financial consequences. Increasingly, we see weak and inept governments up against muscular multinationals. Small state governments find themselves negotiating with companies whose revenue is many times their size. Even in our largest state, New South Wales, farmers are living with the consequences of the former Labor government's decision to issue hundreds of contentious mining leases in a desperate grab for revenue. Australia needs to reform its regulation of the mining industry, in particular by embracing greater co-operation between state and federal governments than our 1901 Constitution provides for. Without such reforms, a handful of multinational companies will continue to profit enormously from resources that by rights belong to all Australians. Under our current system, those most directly affected by mining projects—local communities, regional towns, Indigenous land-owners—often benefit very little.

Politicians and economic experts proclaim that the good times will roll on for decades, transforming the nation into a resources superpower—the Saudi Arabia of the Asia-Pacific region. Surging demand for our dirt and gas has revived the euphoria of the 'rush that never ended'. But this boom should instead remind us of Donald Horne's observation that we are a fortunate country run by second-rate people. We think we are the lucky country, but what we really have is dumb luck—too much luck, more than we know what to do with. Unless we manage this extraordinary boom more effectively, our good fortune will curse future generations.

Bigger, Deeper, Emptier

'The holes will be empty one day and they [the minerals] don't only belong to us; they [also] belong to our successors as residents of this country.'—Roger Corbett, former Woolworths boss and Reserve Bank board member, February 2011[3]

Late one evening in April 1999, Foreign Minister Alexander Downer retired to his hotel room in Bali after attending a special summit with Indonesia's president, B.J. Habibie, on the self-determination ballot to be held in East Timor the following August. After slipping into a dressing gown and slippers, he invited some members of his travelling party to his room, where he lit a pipe and shared a few drinks. As the son of a Menzies cabinet minister and grandson of one of the Australian federation's founding fathers, the plummy-voiced Downer looked and sounded very much like a scion of the Adelaide establishment in which he was raised. On this evening, he could easily have been mistaken for a hereditary peer in the House of Lords.

As the foreign minister held court, the conversation quickly turned to oil. The prospect of an independent East Timor raised issues over control of the Timor Sea's significant oil and gas resources, which had in past decades strongly influenced Australian governments of both hues to support Indonesia's invasion and brutal subjugation of the territory. Would Australia grant an independent East Timor its right to a fair share of the resources, Downer was asked.

The suggestion was treated with disdain. The Timorese, Downer thundered, would 'piss it up against the wall'. A far better outcome for all concerned would see Australia retain control of the resources. In return East Timor would be given foreign aid, operating like an Australian protectorate, he explained.

A decade later, the tables have turned. Downer's own government did exactly what he predicted of the Timorese, while newly independent East Timor (officially known as Timor-Leste, population 1.1 million and one of the poorest countries on earth) has become a stellar example of how to manage resource revenue wisely.

East Timor's leaders knew full well the pitfalls of resource wealth, as some of them had spent many years living in exile in Africa. The resource revenue of these countries had been badly managed, leading to decades-long conflicts and leaving the people impoverished. When East Timor became independent in 2002, its leaders were determined to ensure that the oil fields in the Timor Sea did not doom their people to the same fate. They looked to examples like oil-rich Norway for guidance. Norway, a country of 5 million people, has used its revenue to create a sovereign wealth fund worth almost $600 billion—the second largest such fund in the world. All of the assets are invested in foreign equities and bonds, so that instead of its petro-dollars driving up the krone and damaging other exporters, they drive up the value of the fund.

In 2005, East Timor launched its own petroleum fund for the safe management of oil revenue. Instead of disappearing into consolidated revenue, all tax revenue is diverted into the fund, where it is invested in risk-free government bonds. Overseen by the national parliament, the government can draw on it to finance its annual budget. Each year, the finance ministry calculates how much revenue can be withdrawn without threatening the fund's future sustainability. This figure, known as the 'estimated sustainable income' (ESI), is based on the amount of money in the fund and the net present value of the revenue to be collected from all known resources in production. As the ESI is published publicly and parliament must approve all withdrawals from the fund, public scrutiny is focused on how actual spending compares to the sustainable spending limit. The fund is designed to transform a non-renewable resource—oil—into a financial asset that can be maintained in perpetuity. In the space of six years, tiny East Timor has accumulated more than US$7.7 billion in savings, the equivalent of Australia having saved almost $2 trillion over the same period, based on the relative size of national income. Having established the fund just three years before the GFC, East Timor became one of the few countries in the world not to suffer financial stress as a result of the crisis.

Other resource-rich countries have set up similar wealth funds, often called stabilisation funds, to help with short-term fluctuations in revenue. Stabilisation funds serve to apply a brake to government spending during boom times and put downward pressure on the exchange rate, as governments sell local currency in order to acquire foreign currency assets. Such funds ensure that there is money left over when the boom turns to bust. Over time, they can also be transformed into sovereign wealth funds. Sovereign wealth funds have the longer-term aim of transferring wealth to future generations who will not have the benefit of significant resources. These funds aren't going to solve all of the problems of a resource-rich country; their success depends on how well they are managed and the resolve of the people and government to get it right. Chile has shown how to run an effective stabilisation fund; it was able to build up savings during last decade's mining boom, and then use those savings to bolster its economy during the GFC. Papua New Guinea set up a stabilisation fund in the mid-1970s, not long after it became independent, but the money was later squandered. More recently, however, it has had some success with a policy introduced in 2008 that automatically saves all additional revenue above the long-term average, although it currently invests this entire windfall in infrastructure. In recent years, Australian Treasury economists have advised PNG on how to set up three sovereign wealth funds to manage LNG revenue. Australia, meanwhile, has no such policy in place. It sounds like sheer hypocrisy, and it is.

INHERITING NEW ZEALAND

The mining boom that got underway in the middle of last decade has already made us the richest citizens in Australian history. The extra income amounts to about 15 per cent of our economy, or about $190 billion each year.[4] It is as though we inherited the entire economy of New Zealand in one fell swoop—and if prices remain steady, we will keep inheriting it year in, year out. In fact, with iron-ore exports set to double, and LNG exports on track to quadruple, we'll probably inherit two New Zealands before long. When Australia became a federation in 1901, we missed out on having New Zealand as a seventh state, but the mining boom has made up for it.

In the early 1960s, mining represented around 2 per cent of GDP and 8 per cent of exports. By the end of the last resource boom in the early 1980s it had more than doubled, making up about 6 per cent of our economy and 20 per cent of our exports. The current boom has so far almost doubled mining and energy's share of GDP. Mining now contributes almost 60 per cent of export receipts, worth $165 billion in 2010. These figures don't include the likely doubling of mineral and energy exports over the course of this decade, as production responds to soaring prices.

The boom has vastly increased our buying power, which is measured as the ratio of export to import prices (economists call this the terms of trade). Our terms of trade are now the highest on record, 60 per cent higher than the average for the twentieth century. Think about that: we are 60 per cent better off in purchasing-power terms than our parents, grandparents and great-grandparents, and yet many people still complain about doing it tough. One of our top economic experts, the governor of the Reserve Bank (RBA), Glenn

Stevens, put the increase in our buying power in stark but very simple terms. Five years ago, one shipload of iron ore bought 2,200 energy-intensive flat-screen television sets; today, the same cargo buys 22,000 such TVs. This change has come about in part because television prices have fallen, but it has more to do with the price of iron ore having risen by a factor of six (a few months after this speech, the increase was ten-fold to almost US$190 per metric tonne).[5]

Stevens says that the rise in incomes from the mining boom, combined with investment in even more capacity, makes the current boom one that we may experience only once or twice in a century. Which begs the question: what are our politicians doing to ensure we maximise the benefits from this boom, both for present and future generations? Despite the incredible lift in our fortunes, there is no debate about how best to manage this windfall; there is no debate about how to make sure we get the maximum value out of these finite minerals. The minerals and energy boom is going to shift the centre of gravity of our economy, forcing our society to undergo a wrenching transformation. It could make us very rich for a few years. But as has happened before, capacity may catch up with demand just as prices are falling thanks to a similar supply response around the world, and just as the global economy comes off the boil.

These are serious challenges, but the frenzied pace of the boom allows little time for serious discussion. Our politicians have become so caught up in the boom, so enamoured of being photographed in hard hats and fluoro vests, that they want it to become even bigger. Miners want to rip our resources from the ground as quickly as possible, and the government is more than willing to oblige. To this end, mining companies have secured concessions that allow them to import entire workforces for projects worth more than $2 billion. They now import most of their equipment and send their metal fabrication offshore; projects are assembled like Meccano sets while local businesses miss out.

In January 2011 Martin Ferguson, who is minister for the competing sectors of resources and tourism, unveiled Santos's new coal-seam gas (CSG) project in Gladstone, one of four huge projects that will transform southeast Queensland's economy. As Ferguson triumphantly declared, Australia was on the cusp of emerging as an energy superpower: 'This effectively represents a new industry for Australia: LNG opportunities out of coal-seam methane. From Australia's point of view it is very significant because it further positions us as a reliable supplier of energy products. We are already the major exporter of coal, second or third major exporter of uranium and now potentially the second major exporter of LNG from the west coast, Northern Territory and east coast of Australia by about 2015–16.'

CSG is a new industry that brings with it a raft of long-term environmental risks and uncertainties, especially in regard to its impact on water aquifers and its use of a controversial hydraulic fracking technique that injects chemicals and water into coal seams to release gas. Some of these dangers have been flagged in very strong terms by Australia's water commissioner, Chloe Munro. But in the rush to become a major LNG supplier there has been no time given to consider these proposals properly and carefully. Ferguson's litany of energy export achievements gives the impression that he believes 'too much energy is never enough'. Just months after BP's Gulf of Mexico oil spill, he gave permission for that company to drill for oil in the treacherous waters of the Great Australian Bight. And he has since opened up the Woomera rocket range to mineral exploration, as though we didn't already have enough of the stuff.

'Marn', as he is known in the media, seems to run on the high octane generated by the LNG projects he has been opening. He is a terrific minister: he knows his industry backwards and has a refreshing integrity. He would not have been out of place at the Chifley cabinet table when they drew up the plans for the Snowy Mountains Scheme. But in the resource-boom era, ministers like Ferguson are focusing their enthusiasm on a different kind of engineering feat, one that depletes rather than adds to our national assets. And when it comes to environmental consequences, governments are tending to take a 1940s view of things, disregarding some obvious and potentially damaging side-effects.

When asked if this boom is running too fast, Ferguson says there is no way he'd want to see Australia forego investment opportunities. He curses our loss of 'market share' during the last boom because of skills shortages. Determined not to miss out this time around, recent Australian governments have gone out of their way to create a new resources rush, what Ferguson boastfully describes as the biggest 'investment pipeline' in the world. Some experts, like Dr Stephen Grenville, former deputy governor of the RBA, see it differently, instead characterising the current rush as a 'wild-west resource stampede'.[6]

Mining and energy companies seem to have us fooled with the notion that markets will disappear if we don't exploit our resources as quickly as physically possible. On the contrary, leaving the stuff in the ground will more than likely mean it becomes more valuable, especially our gas in an era of peak oil. Ferguson does acknowledge that this all-out pursuit of a bigger boom puts pressure on our economy, especially on interest rates and on the dollar. 'Do you slow it down and also risk losing investment?' he asks. 'This is the issue that is confronting the government; it is also the issue confronting Treasury and the Reserve Bank. Yes it is challenging us, but we think we can manage it. We have got to be very careful and that's why every board meeting must be an interesting discussion.'[7]

Ask any federal minister or frontbencher about the need to save during the boom and they will reel off a list of immediate priorities and excuses. They'll talk about the 'higher rate of return' we can achieve from investing in infrastructure, or the need to deliver tax cuts. It is true that some investment to address bottlenecks is needed, but big spending on these projects during the boom years often adds to inflationary pressures. If we had a stabilisation fund, we'd save a lot during the boom years and invest a little, and then in the slower years draw on our savings to invest in infrastructure and stimulate the economy. But the problem with our political system is that our leaders are always looking for a quick fix.

GENERATIONAL THEFT

People today should look back with great pride and gratitude at the legacy left by our forebears, who invested heavily in infrastructure and built enormous landmarks that remain a source of wonder to this day. We owe special gratitude to the Diggers who fought in two world wars and then returned to play their role in nation-building. Those generations built our mass transit systems; they built ambitious water and irrigation projects like the Snowy Mountains and Ord River schemes, along with a swag of other dams; and they built icons like the Sydney Harbour Bridge and the Sydney Opera House.

Past generations may not have directly compensated us for exploiting non-renewable resources, but they certainly did pass on a great deal in terms of capital assets. The money that flowed from the Victorian and New South Wales gold-fields built the grand buildings of Melbourne, along with elegant cities like Bendigo and Bathurst and road and rail networks in both states. The same can be said of the Kalgoor-lie goldfields, which proved crucial to the development of the vast state of Western Australia.

Fifty or one hundred years from now, future generations of Australians are unlikely to look back on us in the same way. We enjoy an inflated standard of living largely under-pinned by a new breed of digger who has no interest in leaving a lasting legacy. Data produced especially for this book by the Australian Bureau of Statistics (ABS) show that local, state and federal governments combined are spending about $25 billion *less* per year on public works in today's dollars than in the 1960s and 1970s. At the same time, we are spending twice as much on consumption and social welfare payments. In the 1960s, social assistance transfers amounted to around 4 per cent of GDP; now they have doubled to 8 per cent. Consumption expenditure by these governments has increased from 6 per cent of GDP to 11 per cent of GDP. See the trend? We are spending about twice as much on benefit payments and consumption, and about half as much on fixed capital. Not only are we funding our lifestyle by running down non-renewable resources, we are also leaving much less behind in terms of capital assets.[8]

Without a savings plan to compensate future generations, this mining boom is far worse than people selling off their assets to fund an extravagant lifestyle, thereby leaving nothing for their children. It amounts to theft by the present generation from those in the future. The extra spending might be good for the economy in the short term, but it will leave our children and grandchildren worse off. The scale of this swindle is already breathtaking, and it has only just begun. I happened to mention this idea to a couple of friends recently and they said it reminded them of their sixteen-year-old son—we'll call him Isaac—who progressively sold off his possessions on eBay to finance his busy social life. Admittedly, Isaac sold things he wasn't using, like a surf-board and a drum kit, and his prospects look a lot better than Australia's. He now has a part-time job and can rebuild his assets over the course of his working life. Australia, how-ever, is selling off natural assets that cannot be replaced.

Many Australians probably think our resources will last forever, a view encouraged by historian Geoffrey Blainey in his book *The Rush That Never Ended.* But mineral and energy resources are finite, and we don't have as much of them as we think. The most recent surge in production has cut deeply into known reserves. They say diamonds are a girl's best friend because they last forever. By the time a girl born in 2011 turns twenty, however, Australia will have no diamonds left unless new reserves are found. Manganese resources will also be finished, while gold will be gone by the time she turns thirty, and silver and zinc by the time she turns forty-five, according to estimates from the national geological survey agency, Geoscience Australia (GA). And as this 2011 baby settles into retirement, Australia's supposedly endless supply of iron ore will have been exhausted. The current life expectancy of known iron-ore reserves is seventy years, down sharply from GA's 1998 estimate of 100 years. The end of black coal will not be far off; its life expectancy has come down from 180 years to 100. These figures are based on the information published in the annual reports of listed companies, so they may even overstate the quantities: companies have a financial interest in inflating their reserves.[9]

New deposits are being uncovered all the time, but these gains are being more than offset by the surge in production, which these estimates do not take into account—they are based on *current* production volumes. Mining companies have plans in place to double iron-ore production this decade, which could halve the life expectancy of reserves to around thirty-five years. It is true that we could always dig up more coal, but at what price? Digging up the black soil of the Liverpool Plains in New South Wales would be a huge price to pay to produce more coal to feed the furnaces of China and India. We would sacrifice some of Australia's best farmland, potentially damage water aquifers and add to pressure on scarce water resources. We may well find new deposits, but the days of turning up El Dorados appear to be over. Not since 1975 have mining companies made a major find, what geologists call a 'world-class greenfields' deposit. The last such discovery was BHP's Olympic Dam, according to the geologist Dr Ian Lambert of Geoscience Australia. This find came about when Western Mining Corporation, now owned by BHP, happened upon an enormous copper, gold and uranium deposit while investigating a 'geophysical anomaly'. It seems that most of the easy pickings have now been uncovered; what will be left as the boom progresses is lower grade ore, which sells for a lower price and requires more processing.

When it comes to gas, the US multinational Chevron has tried to convince Australians of the benefits of developing our petroleum resources with full-page newspaper and magazine advertisements. A close look at the data, however, indicates that it might not be in Australia's interest to take advice from energy multinationals, for we are developing these resources at an unsustainable rate. The federal government's 2010 Australian Energy Resource Assessment says current reserves of conventional gas have a life expectancy of sixty-three years, while the estimate for unconventional gas, meaning coal-seam or shale gas, is 100 years.[10] The report admits that these estimates are based on *current production levels.* They don't take into account the export contracts signed in recent years for the Gorgon and Pluto projects on the northwest shelf and for Queensland's new coal-seam gas exports, or rising domestic demand, which together will more than triple total gas production and quadruple LNG exports over the next decade.[11] So those estimates of sixty-three and 100 years should be divided by three or four. Under current sales contracts, gas reserves could be exhausted by the time that girl born in 2011 turns twenty-one. These inflated figures have not only conned the Australian public; they are also quoted throughout government reports, including the Henry tax review. By excluding these new contracts, which have already been signed and are public knowledge, the Australian government is grossly overstating the extent of our gas resources, giving all of us a false sense of security.

The expansion of LNG exports puts Australia on track to become by the middle of this decade the second biggest exporter after Qatar.[12] The scale of new production is breathtaking. LNG exports sourced from Woodside's northwest shelf platforms totalled 16 million tonnes in 2008–09, earning $10 billion. But the Gorgon project alone will double that output, and this and other LNG projects involving investment of around $150 billion could increase Australia's supply by a phenomenal 116 million tonnes—seven times the 2008–09 export level.[13] This expansion is taking place even though Australia's reserves are only ranked *twelfth* in the world.[14] We are trying to punch well above our weight in the LNG stakes, and we are developing new supplies at a time when LNG production is soaring around the world. This bounty will be exhausted much sooner than we think, and it would be more valuable if it were left in the ground for another decade or two at least.

Designing good policies to manage Australia's prosperity is made more difficult by this false complacency. The industrial awakening of China and India has led some to imagine that the business cycle no longer applies. They assure us that we have entered a resources 'super-cycle' that could span decades. With such euphoric predictions, there is little thought given to some of the well-documented pitfalls of resource abundance.

THE RESOURCE CURSE

It is seems inconceivable but it is true—numerous studies all around the world over a long period have found an inverse relationship between natural resource endowments and overall prosperity. For developing countries, resource riches have far too often led to corruption, cronyism and deadly conflict. But corruption and conflict are only part of the explanation. There's a broader effect that takes place as resource wealth drives the local currency up and sucks the life out of the rest of the economy. When the resource boom ends, these countries find that the non-resource economy is unable to get back on its feet.

Such riches to rags stories have given rise to terms like 'the resource curse' and 'the paradox of plenty'. In 1976, Professor Bob Gregory of the Australian National University observed the shrinking of the non-mining sector during the 1970s resource boom (the one made infamous by the spectacular Poseidon bubble). Gregory concluded that mining's impact on manufacturing had been just as great as the 25 per cent tariff cut. His observations became known as the 'Gregory Effect'. He wrote:

> The rapid growth of Australian mineral exports, through its effect on the balance of payments, is a significant force for structural change in other sectors. From the viewpoint of the rural sector which exports and the manufacturing sector which competes with imports, this force will be similar to that which would flow from very large tariff changes.[15]

At the same time, the Netherlands faced a similar shock. The country was being inundated with petrodollars from its share of the North Sea oil, leading to a high exchange rate and higher costs for manufacturing and many other export businesses, eventually causing them to shrink or go bust. The *Economist* magazine coined the term 'Dutch disease' in 1977. The 'Gregory Effect' was not nearly as catchy, but it had the same meaning, and Gregory observed these changes first.

When governments misuse revenue, they help to spread the resource curse. Instead of investing sensibly, many resource-rich nations spend money on expensive military hardware and grandiose edifices. This is very common in the developing world. Africans now joke that they have a new wild animal: the white elephant. In oil-rich Angola, the government has recently built twenty-four new hospitals, but it cannot find medical staff; the country has only 1,500 doctors. Elsewhere in Africa defence spending is rising sharply while it has fallen around the world in the post-Cold War era.[16]

A global study by Professor Paul Stevens of twenty-nine resource-rich countries found that government spending is one of the key drivers of the resource curse. The negative effects 'may be related to economic or political flaws in managing natural resource revenue'. Stevens found that only four developing countries—Chile, Malaysia, Indonesia and Botswana—had succeeded in extracting wealth from the ground without succumbing to these pitfalls. This is an appalling record, and it shows that without strong policies, the more resource wealth a country has, the less likely it is to succeed.[17] Two particularly spectacular downfalls have occurred in Zambia and the phosphate-rich island of Nauru.

Formerly known as Northern Rhodesia, Zambia became instantly wealthy in the 1960s thanks to its extensive copper riches. The place was touted as being on the cusp of joining the first world and Zambians believed this was their destiny. Yet the country quickly went from being an exporter of agricultural produce with a diversified economy to being a net importer of food. As James Ferguson observed in his book *Expectations of Modernity*, everyone believed the country was destined for greatness. Zambian miners were so rich that they wore suits on weekends that were tailored in London.

The idea that Zambia was destined to move ahead to join the ranks of modern nations, and that 'development' would lead Zambia to ever-greater urbanisation, modernity and prosperity, had come by the 1960s to be accepted both by academics and national and international policy-makers, and by a wide range of ordinary Zambians as well.[18]

Zambia was taking an extreme risk by relying on copper for most of its export income, and when the global recession of the 1970s punctured the bubble in copper prices it demolished the country's standard of living. In the space of fifteen years, the buying power of Zambia's exports fell by 87 per cent. By the end of the 1970s, infant mortality had soared 35 per cent and life expectancy had fallen by five years to fifty.[19] Had Zambia and its advisers not been so over-confident, the country would have built up savings during the boom so that it could maintain or at least buffet its living standards during the bust. But the country's leaders had behaved as though the business cycle no longer applied and it had no parachute for the inevitable crash.

Closer to home there is the case of Nauru, whose people lived a first-world existence for a few decades. The island state owned a now-defunct airline, it owned an office tower in Melbourne named Nauru House, it bought into a theatre production in London and its people developed a taste for imported French caviar.[20] But now that its phosphate bounty has been almost entirely exhausted, the country is a physical and financial wreck, dependent on foreign aid and the hope of becoming a processing centre for asylum seekers.

In Australia's case, the ill-effects of booms tend to be channelled through the exchange rate, as shown by the Australian dollar's surge above parity with the US. But this is not just because we are selling more of our minerals. The resource boom has transformed the 'Pacific peso' into the new hard currency of the region, especially in the wake of the US dollar's demise, which creates a virtuous (or vicious) cycle as global portfolio investors park their money in Australian assets. Wealthy people in Asia now choose to hold Australian dollars rather than greenbacks. The high dollar reduces returns to local businesses, most of which are not benefiting from the very high prices of the resource sector. As imports become cheaper, local business will struggle to compete. A research paper presented to the RBA board in 2010 predicted that Australia could see an additional $120 billion in portfolio investment each year, putting further upward pressure on the dollar. RBA economist Katrina Clifton argued that putting assets into foreign currency via a stabilisation fund was the best way of mitigating these effects.[21]

The resource boom last decade inspired some truly reckless spending by the Howard government, and this has continued under Labor. The stimulus spending during the GFC was beneficial for the economy, but the school building and insulation schemes set dangerous precedents. Labor's latest cash splash gives the elderly $300 each to pay for digital set-top boxes that can be bought and installed for a third of that amount. The grand prize for fiscal recklessness goes to the $36 billion National Broadband Network (NBN), a project introduced without a cost-benefit analysis. The NBN delivers lightening-fast internet via optical fibre to nearly all Australian households, even though there is no evidence they've demanded such a service. If demand were to emerge, it's a project that would be better undertaken by the private sector; giving a single program like the NBN a monopoly will stymie competition and innovation.

As the tax revenue from the mining boom grows in coming years, so will the imagination of our politicians for even more grandiose spending under the guise of 'nation building'. What's all the more galling is that this enormous amount of money is being devoted to spurious projects when communities are crying out for better roads—especially in mining areas, where they are choked with heavy vehicles—and better public transport. While the NBN is being rolled out, state and federal governments have largely ignored community calls to fix deadly black spots on the highways that feed the Queensland mining centres around Mackay, Gladstone and Toowoomba. The Peak Downs Highway west of Mackay has become a dangerous thoroughfare, with a constant stream of heavy vehicles and commuting miners heading in and out of the Bowen Basin coal mines. This is a problem caused directly by the mining boom—and yet the governments so eager to promise money elsewhere have been slow and deficient in responding.[22]

Australians should pay close attention to the experiences of Zambia, Nauru and other victims of the resource curse. While it might sound improbable, our governments are displaying similar myopia and we are in danger of repeating their mistakes.

Top Gear, Second and Reverse

'There's never been a better time to visit the United States.'
—Queensland Treasurer Andrew Fraser on how the soaring Australian dollar has flattened the state's tourism industry, March 2011[23]

One of the fundamental ideas behind Australia's modern economy is that of 'comparative advantage', which has been applied by many of our economic experts with religious zeal over the last three decades. According to this principle, countries are better off making their economies more specialised, focusing on those industries in which they are most globally competitive, rather than attempting to maintain a broader base. This notion has driven the removal of protection and the demise of manufacturing in Australia, and is now behind the stampede into the resource sector. We are told that supplying low-cost energy and minerals is what we do best, and that it therefore makes sense to make this our focus.

The theory of comparative advantage was first developed by David Ricardo, a nineteenth-century English economist who was a passionate advocate of the benefits of international trade. Ricardo's theory claimed that all nations can benefit from free trade, even if some countries lack absolute advantage. The theory has underpinned the global trading system and has proved beneficial when it has not been taken to an extreme. In his *Principles of Political Economy and Taxation*, first published in 1817, Ricardo wrote:

> Under a system of perfectly free commerce, each country naturally devotes its capital and labour to such employments as are most beneficial to each. This pursuit of individual advantage is admirably connected with the universal good of the whole. It is this principle which determines that wine shall be made in Portugal, that corn shall be grown in America and Poland, and that hardware and other goods shall be manufactured in England.

To demonstrate the benefits of free trade and specialisation, Ricardo offered the examples of Portugal and Britain. He argued that Britain's most efficient industry was cloth making and Portugal's was wine. Rather than the two countries each making both goods, the welfare of both would improve if they specialised in what they were best at. Even if Portugal could produce cloth more efficiently than Britain, it would still be better off specialising in wine (in which it had an even greater relative advantage), exporting it to Britain and using the revenue to purchase British cloth.[24]

This theory holds that a country should use its resources as productively as possible, which is common sense. Australia's removal of the tariff wall led to the wholesale shutting down of uncompetitive manufacturing, but new service industries like tourism and education blossomed with the benefit of a lower Australian dollar. The floating of the dollar and the abolition of tariffs went hand in hand. Many people would agree that we are better off as a result. It made no sense for Australia to have effective tariffs on clothing of more than 100 per cent when this generated only a few thousand jobs and imposed higher costs on everyone. With this protectionism out of the way, gone are the days when families spent a small fortune clothing their children.

Now that the resource boom is in full flight, our politicians and our experts have been chanting comparative advantage like an article of faith. Glenn Stevens has told us that the slow lanes of the economy will have to accept their inevitable decline. 'Ultimately, it looks likely that the mining sector and the areas that supply it will grow, and some other industries will, relatively, get smaller. And at this point, much of the impact of the recent resource price changes is yet to be seen,' Stevens told a conference in the rural town of Shepparton in late 2010, where the local farmers were no doubt feeling the squeeze from the higher dollar.[25] A former head of Treasury, Ken Henry, has also said that in order to benefit fully from the resource boom we will have to shift resources into the fast lane, leading to the relative or even absolute decline of less efficient sectors like manufacturing.

But there is one glaring weakness with this theory, which our economic experts never discuss. Comparative advantage involves narrowing a country's economic base until we eventually rely on just a few industries, or even one main industry, for export income. It is like investors who decide they want to put everything into the resource sector, rather than having a more balanced portfolio spread across different industries and businesses. Think about how you would like to see your superannuation invested. Would you feel comfortable if your fund manager decided to shift most or all of your assets into resources? In the case of Portugal, what would have happened if the country had opted to specialise in wine at a time when viticulture production was ramping up around the world and other countries were entering the export market, thereby lowering prices? Portugal would have sacrificed its competitive advantage in the emerging cloth industry on the altar of 'Ricardian efficiency'. Australia is doing the same as we specialise in resources at the expense of industries like export tourism and education, which have been painstakingly built from scratch in the post-tariff world.

For Australia, one profound consequence is that the more China and India industrialise and demand our raw materials, the more our economy resembles that of a developing country. This observation was made by Stanford University's Professor Paul Ehrlich during a 2009 visit to Australia. By becoming the quarry to these countries, Australia is exploiting its 'comparative advantage' as a low-cost producer of minerals and energy, instead of maintaining a broad-based export sector.[26]

Most economists in Australia are loyal Ricardians who say we have nothing to fear from exploiting our comparative advantage to the max as we ride the resource boom. They've come up with a neat way of conveying this to ordinary Australians: it's the 'two-speed economy', a milder way of describing Dutch disease. It means that the resource sector races ahead and draws labour and capital from other sectors, which decline in relative or absolute terms. Others say it is more like a three-speed economy: the turbo-charged minerals sector shifts into top gear; sectors not exposed to global competition like services are in second; and sectors in competition with offshore rivals, like manufacturing, tourism and education, get squeezed by the high dollar and are shunted into the slow lane or even go backwards. Adhering to the dogma of comparative advantage, the mining industry and our economic experts want to see investment and labour shift quickly and seamlessly from the slow-growth economy to the high-growth sectors, even though this means we will be more vulnerable to anything going seriously wrong in the global economy.

This process affects just about every aspect of our daily lives. The investment surge explains why we have a housing shortage in the eastern states when interest rates are relatively low, population growth is strong and investors can even write off the cost of borrowing. A Treasury briefing paper prepared for Treasurer Wayne Swan and obtained under FOI concedes that the mining boom is responsible for this conundrum:

> It is certainly true that much of the 'heat' in the housing market in recent years has been reflected in prices, with investment activity remaining subdued. But this has also been a period where strong profits in the mining sector have fuelled an investment boom that has drawn resources away from sectors such as housing.[27]

Mining booms put added pressure on interest rates because of the sector's intense demand for capital, which discourages investment in other capital-hungry sectors like housing. Similar effects are being felt when it comes to skilled labour. In the Pilbara, the epicentre of the boom, the demand for labour is so intense that mining companies can't find local businesses to meet even basic needs; they have been known to fly bread and sandwiches 2,300 kilometres from Perth, and to fly laundry back and forth as well. The town of Karratha, centre of Woodside's main LNG operation and Rio's export hub, has no butcher or bakery. Chris Adams, who runs the Western Australian government's Pilbara Cities office, says mining companies are like vacuum cleaners: 'They suck everything up.'[28] The defence force has felt these pressures acutely. The Royal Australian Navy can now keep only a handful of its ships and submarines operational, in part because it faces difficulty recruiting and retaining personnel as a result of competition from mining companies. The RAN requires technical people who have exactly the skills demanded by the miners, but it can't pay $200,000-plus salaries. A former defence minister accused mining companies of sending recruitment staff to hover around bases in a bid to poach skilled personnel.[29]

The mining boom is making growth in our economy stronger and thereby lowering unemployment across the board, but parts of the country are being harmed by the soaring dollar and the resource sector's intense demand for capital and labour.

TUMULTUOUS TIMES AHEAD

Commodities are an inherently unstable business. Over the decades, they have tended to have a few bumper years followed by long periods of over-supply. This short boom and long bust cycle has been an inherent part of Australia's history ever since the first mining boom began in the mid-1800s. Data compiled by the RBA economists John O'Connor and David Orsmond show that base metals prices were largely flat between the 1920s and the mid-1960s, until the Vietnam War. From the 1970s onwards there was a steady though volatile downward decline, until the spectacular reversal last decade. A similar pattern is evident for oil, coal and gold, with a depressing trend line interrupted by occasional peaks.[30]

The risks that now lie ahead could well be more severe than the near-death experience the global economy faced in 2008, warns Professor Warwick McKibbin, who has served on the RBA board for a decade. He says loose liquidity has driven commodity prices sky-high, creating an even bigger bubble than the one that preceded the GFC. Professor Nouriel Roubini, an economist who predicted the GFC, says very simply that China's economy is a bubble. Roubini argues that the country's economic boom relies excessively on debt-fuelled investment. New evidence shows that this surge is creating ghost cities populated by apartment buildings that no-one can afford to buy. Investment accounted for almost 60 per cent of Chinese growth in the five years to 2010, compared with about 25 per cent in the five years to 2000. Investment cannot keep endlessly expanding; as with a Ponzi scheme, when new contributions dry up, the whole thing comes crashing down.[31] A similar thing happened in many Asian countries before the 1997 financial crisis; those economies imploded when the banks stopped lending. The sharp correction that Roubini predicts for China within the next two years means that the high prices used to justify the massive investment in new capacity in Australia will no longer prevail. Even if the Chinese bubble doesn't burst, commodity prices are likely to suffer a correction given the new supplies expected to enter the global market.

If Australia were a business, it is hard to imagine that our board of directors would be putting more than half their investment capital into an industry with such a poor track record and myriad harmful effects on other branches of the company. But that is exactly what our directors are doing. Without the restraint of more effective taxation, our directors are displaying unbridled exuberance. An April 2011 survey by ABARES, Australia's natural resource advisory body (previously known as ABARE), identified ninety-four mineral and energy projects at an advanced stage of development, with record capital expenditure of $173.5 billion, plus another $256 billion in projects awaiting a final investment decision.[32] In 2011–12, more than 70 per cent of all investment in Australia is slated for the minerals and energy sector, according to the ABS capital expenditure survey. An industry that makes up just one-tenth of the economy now commands more capital than the other 90 per cent.

Brian Fisher, a resource economist who headed ABARE for many years, supports the principle of pursuing comparative advantage but says our economy will be exposed to a more tumultuous world. 'I think we are headed for a world where there will be much more volatility, periods of high prices and periods of low prices,' he says. 'There could be strong surges of growth, followed by macroeconomic instability in the developing world that cascades back onto the developed world, with big swings in prices.'

THE END OF INNOVATION

In this era of economic effervescence, it is increasingly rare to hear business and political leaders talk of innovation as the key to our prosperity; the boom has not only sapped investment dollars from new business ventures, it has also changed our mindset. Gone is the vision of the late John Button, the industry minister in the 1980s and early 1990s who championed innovation in the post-tariff world. The Gillard government's decision in January 2011 to abruptly cut the $1.3 billion Green Car Innovation Fund, which was to offer one dollar of government support for every three dollars invested by car makers, would have Button, the architect of our post-protectionist car industry, turning in his grave. The $800 million saving was made at very short notice and after Holden had committed to build a new small car in Australia. The decision reflects the new thinking: now that we are awash with mineral wealth, we don't need manufacturing.

Venture capital is also being starved of funds, leading to an exodus of start-up technology companies to foreign shores. Dr Katherine Woodthorpe, the chief executive of the private equity and venture capital association AVCAL, says the government has in recent years become less focused on the need to drive innovation in the economy. She cites the decision to slash funding for commercialisation of research and development. Abolished in 2008 to save $700 million, the Commercial Ready grants scheme that was cut was first developed (as Industry Research and Development, or GIRD, grants) by John Button in the 1980s. Asked if the mining boom had made the government apathetic about innovation, Woodthorpe says, 'It's a symptom. If we were serious about not becoming just a quarry, we'd be putting more effort into commercialising the $7 billion we invest in public sector R&D such that it can deliver economic gains. At the moment that's not happening. We are not an economy that turns R&D into start-up businesses.'[33]

KAKADU OR COAL?

Relying on resource commodities to pay your way in the world thus makes countries more vulnerable to global prices and supply responses. As Warren Buffett has said, there ain't anything special about the stuff—it has no unique 'franchise'—and that's why the world's most successful investor avoids the sector. Coal and iron ore are nothing like a Great Barrier Reef or Kakadu holiday experience, which are unique in the world and which European and Asian tourists are willing to pay very good money for. They are nothing like our grain, beef or dairy exports, which benefit in global markets from our reputation for a healthy food production chain.

They are nothing like the education services that were earning Australia close to $20 billion a year until recently. And nor are they anything like the medical products made by CSL, Cochlear or the emerging adult stem-cell company Mesoblast. These products are either unique or have brand value that is difficult or impossible to copy and which has therefore secured them a place in the global economy.

But while commodity prices are high, the boom will play havoc with these other exporters, who will see their foreign currency earnings slashed when they convert them into Australian dollars. The soaring demand for minerals has driven the dollar to parity with the US currency and beyond. This is fine for miners, as they are compensated by astronomical prices. But other exporters are being crunched, and they will continue to be crunched, even though our economic experts tell us that we have nothing to fear.

Treasury's chief economist, Dr David Gruen, admits that the parts of the economy not linked to resources 'are currently facing severe and sustained competitive pressure from foreign competitors because of the appreciation of the real exchange rate', which sounds a lot like Dutch disease with an Australian accent. But Gruen then alludes to an invisible hand working its way through the global economy, providing 'rising demand for a range of Australian goods and services—whether they are a range of foodstuffs, Australian tourist destinations, or educational, financial and other professional services ...' No evidence is given to support this claim—the facts are just asserted and we are expected to accept them. Gruen says that it is 'not possible to predict with any accuracy which of these Australian economic sectors will benefit most from the re-emergence of China and India' but in any case, he says, this is still an 'argument for maintaining, and enhancing where possible, the flexibility of the Australian economy'.[34]

Martin Ferguson believes that the emerging middle class in China will provide strong growth for tourism, notwithstanding the impact of the high dollar. He cites an 8 per cent growth in tourism arrivals from China of late, but concedes that the challenge will be how to ensure the sector is still standing when the resource boom abates. 'You have to work on the basis there will be a dip in resources at a point. How do you keep sectors like tourism, the financial sector, all those other associated sectors going?'

It is an excellent question. With the Australian dollar already above US$1, and heading much higher as the boom gathers momentum, Australia's other exporters are proving to be uncompetitive. Chinese people may be more cashed up, but it is very unlikely they will continue to send their children to study here, or come here for a holiday, when the exchange rate is so high. Despite the optimism of Gruen and Ferguson, the evidence so far shows that our export base is becoming narrower.

Australia's two big service exports are tourism and education, and neither of them is doing well. To really understand the impact of the resource boom, Queensland provides a useful case study. It has two competing export sectors going head to head—minerals and tourism. Despite being regarded as a resource powerhouse, economic growth in Queensland has been the slowest of all the Australian states, even before it was hit by floods and Cyclone Yasi. There's a very good reason why this has been the case. Queensland's economy depends on overseas visitors to the resorts that run from the Gold Coast all the way up to Cairns. The high dollar has cut tourism numbers sharply.

The Queensland treasurer, Andrew Fraser, says that slow growth and high unemployment reflect the fact that tourism has been crunched, first by the GFC and then by the high dollar. Because of the distances involved, Fraser notes, most overseas visitors to Australia are very sensitive to price changes. He makes a sad admission for someone from the sunshine state—it is now more attractive for global tourists and Australians to visit the United States. In the wake of the natural disasters the Queensland government told other Australians that the best way they could help out was by spending their next holiday in the state. But this plea appears to have been ignored. Inbound and outbound tourism numbers confirm that the mining boom has not only discouraged international visitors, but has also prompted more Australians to holiday overseas. A report published by the federal Department of Resources, Energy and Tourism includes a graph illustrating that as the Australian dollar rose above US80c, outbound tourists exceeded inbound tourists for the first time since 1987.[35] This trend accelerated as the dollar rose above parity with the US. The latest ABS figures show that Australians leaving the country on short-term trips now number around 610,000 each month, exceeding the number of short-term tourists coming here by 120,000.[36]

The effect on Queensland has been devastating. Home-buyers have experienced negative equity, and now tourism businesses are hitting the wall. Jason Anderson, a senior economist with the property consultancy MacroPlan, says that while central Queensland was booming with the coal-seam gas rush, tourism centres on the coast were going backwards. Cairns and the Gold Coast have some of the highest unemployment in the country, with an average of 10 per cent and 7 per cent respectively.[37] Pockets of the Gold Coast have double-digit unemployment and are experiencing increased violent crime. House prices in these centres fell by around 5 per cent in 2010 and could fall by the same margin again in 2011. Gold Coast developers have begun offering discounts of 10 to 15 per cent on apartments. An unprecedented 15 percentage-point gap has now emerged between the growth of prices in fast-lane places like central Queensland and the depressed tourism centres, Anderson says.[38]

The Minerals Council of Australia (MCA), which is funded mainly by the big miners, has strongly denied that the mining boom is having a negative impact on industries such as tourism. To bolster its case, the MCA commissioned a study by the accounting firm Deloitte, which concluded that the theory of the two-speed economy just didn't stack up. The study said:

> In a modern, dynamic and growing economy, there are always sectors that are expanding and contracting as demand and supply conditions change and prices adjust. Australia does not have a 'two-speed economy',

it has thousands of industries operating at different speeds, with price and resource adjustments taking place constantly.

Seizing on the paper, MCA chief executive Mitch Hooke argued that wide variations in rates of growth in different states prove this point. The recent growth figures showed that while resource-rich Western Australia was charging ahead at 6.5 per cent, states like New South Wales and Victoria, which are not dominated by resources, were also growing at above-average rates. But Queensland, one of our so-called resource juggernauts, experienced the slowest growth in the country, even slower than Tasmania and South Australia. Hooke couldn't contain himself. 'That's right,' he declared, 'Queensland, the mining-rich resource powerhouse, was stone motherless last. According to Canberra's two-speed thesis, Queensland should be roaring ahead of states such as Victoria, South Australia and certainly Tasmania.'[39]

Despite Hooke's bombast, however, slow growth in Queensland actually proves the two-speed economy theory. In the sunshine state, mining is clearly killing tourism, and other export industries as well. As the dollar soars, tourism resorts have been quietly folding. In the first half of 2011 the Couran Cove resort on the Gold Coast and Peppers Bale resort at Port Douglas were two of the headline collapses.

The high dollar is also hitting foreign students looking to study here, a business which in 2009 generated $18.6 billion in earnings. This was before Chris Evans, as immigration minister, tightened up the application process for student visas, and before the high dollar took its toll. Why come to Australia when a more prestigious school in the US, Canada or the UK is now far cheaper? Evans, who has since become higher education minister, acknowledges the dollar is adding to the troubled sector's woes. 'Quite frankly, with the Australian dollar at over a dollar American, that's why we're seeing difficulty. It is cheaper now to go to America and Britain than it is to come to Australia,' he says.[40] Competitors in these countries have seized on this opportunity and opened their doors to foreign students, launching slick marketing campaigns that have outshone Australia's. Chinese enrolments, which had been generating $5 billion a year, dropped 50 per cent after our dollar reached parity with the US dollar. Wang Wei, general manager of Crossworld International Education Centre, an advisory firm for Chinese students, agrees that the dollar was a key factor. 'With [the] Australian dollar solid, the cost of education in Australia is equal to that of the US, if not higher. The US is more attractive to Chinese as a country and American universities are better known to Chinese than Australian ones,' Ms Wang explains.[41] Tourism and education were once stand-out examples of how Australia could diversify its economy in the post-tariff world, but no more.

DIGGING DEEPER INTO DEBT

The rush into resources has consequences for the economy as a whole. Ultimately, increased investment draws on savings, and if we don't have enough savings we have to borrow from elsewhere. This increases Australia's annual trade and income shortfall, known as the current account deficit. Over the past three decades this deficit has averaged 4 per cent of our GDP, but it kicked up to 5 to 6 per cent last decade as the mining boom got underway. It has come down again now that households have learned from the GFC and reined in their debts. But the miners haven't learned the same lesson, and their ambitious investment plans are likely to reverse this improvement. Experts like Warwick McKibbin predict a deficit in the order of 10 per cent of our economy should resource companies go overboard with their investments.

What really worries a former central banker like Stephen Grenville is that this investment and commensurate increase in our overseas borrowing will give foreign lenders greater control over our economy. Grenville should know, as he worked on cleaning up the mess left by the Asian financial crisis in the late 1990s, when overseas lenders cut and ran from debt-dependent regional countries and caused their economies to implode.

> To accommodate the mining boom, we are encouraged to look with equanimity on a widening current account deficit. But there are consequences. It means selling off more of the farm: more foreign ownership. It means we will have a less diversified economy, making us more vulnerable the next time resource prices collapse. And it means increasing reliance on foreign funders to tide us over this adjustment.

Grenville says the GFC shows that Australia should not assume New York bankers will always fund our deficit. Australia only 'scraped through' the GFC, and it did so by requiring the federal government to put its AAA credit rating on the line 'to keep our banks borrowing in New York'. This view is supported by the National Institute for Economic and Industry Research (NIEIR), a private research firm, which has warned that Australia is 'at risk of an Iceland/Ireland type meltdown if there is a sudden loss of confidence in the currency'.[42] The lesson is that if an industry worth just 10 per cent of our economy loses its head and gambles on the resource super-cycle, it has consequences for all of us, especially when it involves making our economy more dependent on foreign lenders. But most of Australia's top economic advisers take a more sanguine view; they have largely dismissed the idea that this boom could end up being a curse. They believe that despite the woeful performance of resource-dependent economies last century, mining and energy in the twenty-first century are a sure bet. On what do they base such optimism?

CITIES AND STEEL

The dawn of what is now called the Asian Century has only just begun. In the space of twenty years, the share of global income earned by China and India has doubled from one-tenth to one-fifth, and in another twenty years their share could reach one-third, according to analysis presented in the 2011 federal budget. By the end of this decade, Asia is expected to have a bigger middle class than the rest of the world combined, and China's will surpass that of the United States. A decade later, two-thirds of the world's middle-class people could be living in the Asia-Pacific region.[43]

The pace of industrial and urban development in China, India and other Asian countries indicates that they will be demanding even more of our mineral and energy exports, although there are likely to be many ups and downs along the way. The amount of steel consumed by these emerging economies is a key reason for the exuberance among our top economic experts. As countries industrialise they consume an enormous amount of coal and iron ore to make steel. Steel is not only used in cars, white goods and electronics; it is also needed to build roads, bridges and high-rise buildings. The consumption of steel goes hand in hand with overall demand for the mineral and energy commodities that Australia produces.

Back in the 1850s, the United States consumed about 8 tonnes of steel for every $1 million in national output, but consumption increased five-fold through the course of its industrialisation, before declining as urbanisation slowed and manufacturing declined as a share of the economy. Japan showed the same trend from the 1950s onwards. When its income per head was about $3,000, steel consumption was just over 20 tonnes per $1 million of GDP, but it then increased three-fold.[44]

The same trends are emerging in China and India, but this time we are dealing with even more intensive steel consumption and a transformation that involves 2.5 billion people.

Since the beginning of China's economic awakening in the 1980s, its urban population has doubled to 47 per cent, and its steel consumption has doubled to 60 tonnes per $1 million of output. But its urban population is still growing and per capita income is less than $5,000. As incomes rise, more people will live in cities and they will buy cars and other durables. In the space of a decade, car ownership has increased from 5 million to 35 million, and it is still rising exponentially. India's steel consumption has shown a much smaller rise to about 16 tonnes for every $1 million of GDP, but its per capita income is still around $3,000 and only 28 per cent of its population live in urban areas. India's 1.2 billion aspiring people have a lot of catching up to do.

These trends indicate that China and India will be buying a lot more of our Pilbara iron ore, Queensland coking coal and Groote Eylandt manganese, along with our LNG and thermal coal for power generation. In 2010, total coal exports were $43 billion, compared to $47 billion for iron ore. Iron ore is expanding rapidly and is on track to far exceed coal as producers in Western Australia put in place ambitious expansion plans. BHP has plans to triple production to 350 million tonnes per annum (mtpa) by the end of this decade, Rio is gearing up to more than double output from 220 mtpa to as much as 500 mtpa, and new entrant Fortescue Metals is trying to triple production to 150 million tonnes. The Chinese state company Sinosteel is negotiating with landowners for a 300-million mtpa operation. Remember, in 2011 Australia's iron-ore output was just over 400 million tonnes, so these increases could easily push annual output to more than a billion tonnes. Coal and iron-ore exports are already on track to top $100 billion soon, or about 8 per cent of our economy, and with these expansion plans those earnings will keep growing—provided, that is, that the ten-fold increase in iron-ore prices over the course of this boom doesn't evaporate as the new supply comes on stream.[45]

THE SUPPLY SIDE

To explain the origins of this boom, most people have focused exclusively on Chinese hunger for raw materials, but this boom is as much a story of weak supply capacity as it is about strong demand. Brian Fisher says that his economists offered regular warnings about poor investment in capacity, particularly in iron ore. 'At ABARE we said there had been underinvestment in iron ore for many years. And then prices went high because there had not been enough investment; supply wasn't keeping up,' he says. Companies had been reluctant to invest because they'd been dealing with the legacy of over-supply since the early 1990s. Now that prices have stayed mad for a few years, the supply response is finally starting to gain momentum. But Fisher worries that by the time the new capacity is ready we may have missed the boat, leading to even more global economic instability. He says companies invariable underestimate the time-lags involved in getting projects approved and pulling together all the components. 'There seems also to be a general tendency to underestimate the difficulty in bringing on supply when markets are tight,' he observes.[46]

The view that Australian companies are over-reacting to spiking prices is shared by the man who will most likely increase interest rates in response to these pressures. Glenn Stevens suggests that much of the new investment is predicated on current price levels remaining high. The Australian economy could be undergoing a painful and costly restructuring—for nothing.

> It would not make sense for there to be a big increase in investment in the sorts of resource extraction activities that could be profitable only at temporarily very high prices. Moreover, the economic restructuring that would reduce the size of other sectors that would be quite viable at 'normal' relative prices and a 'normal' exchange rate—assuming there is such a thing—would be wasteful if significant costs are associated with that change only to find that further large costs are incurred to change back after the resource boom ends.[47]

A drop in prices won't arise only if the global economy hits a speed bump or crashes. It will also come about as other countries respond to the high prices. We hear a lot of people, including our so-called economic gurus, marvel at our abundant resources, as though no-one else in the world can produce coal, iron ore and gas the way we do. Fisher thinks that many people are underestimating the strength of the supply response from other resource economies. 'The current high prices are an enormous incentive to bring more of this resource on line,' he says.

The LNG market is particularly volatile and fast-changing, and Australia's competitors have been busy ramping up production. Demand is underpinned by Asian customers and possibly the effects of Japan's nuclear disaster, but supply is increasing very quickly. The International Energy Agency (IEA), which was created during the OPEC oil shock to stabilise global energy supply, predicted in November 2010 that the 'gas glut' would last another decade. A June 2011 IEA report, however, sees this oversupply ending by 2015 as rising demand heralds a 'golden age of gas'.[48] But supply always catches up and often overshoots. As Australia sinks more than $150 billion into new LNG projects, the Chinese companies CNOOC and PetroChina have taken stakes in US projects to ramp up exports from that country, according to the oil and gas consultancy Wood MacKenzie. The consultancy argued that these new supplies would undercut Australia's export prices by around 10 per cent.[49]

Qatar, meanwhile, not only trumped Australia's bid to host the 2022 soccer World Cup; this tiny Arab state, population 1 million, is also beating us hands down in capturing the LNG market. In the space of a few years the country has increased supply three-fold to 77 million tonnes per annum, five times Australia's production. An economist who has been working for the Qatar government for several years (who asked not to be named) has witnessed the rapid development of the country's LNG business in a way that contrasts sharply with Australia. The Qatar government has maintained majority interests in all infrastructure projects, including LNG production plants, the ships that transport the LNG and some of the re-gasification plants in the countries that buy the gas. The Qatar government therefore captures a majority of the value generated across the entire supply chain, instead of just the tax revenue from production, this economist observes. The Qatar government has built enormous spare capacity—which has contributed to global over-supply—but this means it can capture the lucrative spot market, where buyers can secure immediate supply, as well as long-term contracts. 'They can meet the spot market very quickly because they control the production, the ships and re-gasification facilities,' he says.

There is clear evidence that such state-owned companies have delivered benefits to major resource economies. Norway developed its resources with the state oil company Statoil before partially floating it a decade ago. The government retains a two-thirds interest in the business. With a market capitalisation of $75 billion and operations in thirty-four countries, Statoil is a top 100 global company, one of the world's biggest oil operations and more than double the size of Woodside. And remember, this is a business owned by a country with a population about the same as New Zealand. Australia has never had a debate about creating a state resource company, although this path was advocated by Ross Garnaut and Anthony Clunies-Ross in the 1980s when Australia was first developing its LNG industry. They argued in 1983 that it was 'worthwhile' for governments to invest in such projects because the cost of raising the large amount of investment capital required was lower for governments than for the private sector. Such views, however, soon became unfashionable.[50] Instead, we leave the development of our natural resources to foreign corporations, which control more than 80 per cent of mineral and energy production in Australia. While Qatar profits from the entire production chain, we're content to be compensated with a modest amount of tax revenue and increasingly less in the way of local economic return.

African countries will also compete with Australian producers in coming years. China has been courting resource-rich countries in Africa to obtain access to their mines and oil fields. From Lesotho to Egypt, Chinese construction companies have been advancing Beijing's soft diplomacy by building roads and other infrastructure. China is now South Africa's largest trading partner. It has invested in Mozambican coal. Its state oil company, CNOOC, has been negotiating to secure one-sixth of Nigeria's proven oil reserves in a deal worth up to $50 billion. And the Chinese have also begun tying up supplies of uranium from Niger.[51] Closer to home, they have been very active in oil-rich East Timor, having built a number of palatial buildings for the government, including the presidential palace, the foreign ministry headquarters and the defence headquarters. The Chinese see this as a long-term investment to help secure some of the country's riches. Even New Zealand, which promotes itself as '100% Pure', has been getting into the minerals and energy supply chain, allowing mining in some national parks. It has also turned up evidence of significant oil wealth.

This strong supply response is happening all around the world and will inevitably force down prices, as has happened in previous booms. This will in turn reduce earnings on the hundreds of billions of dollars now been invested in new projects and expanded capacity in Australia. Should boom turn to bust, we will find that many of these new projects are unviable. And by then, our pursuit of comparative advantage may have left us with no other legs to stand on.

We'll Spend It While We Can

'It's the Maserati of public policy'—Shadow Treasurer Joe
Hockey, dismissing calls for a special fund to save windfall resource revenue, February 2011

Mining has been a powerful and positive force in Australia's
development since the gold rush of the mid-1800s. It spurred
the building of towns, railways and roads that opened up
regional and remote areas, but it has also been a volatile sector, inducing periods of economic giddiness that have all
ended with an enormous thud.

Everywhere mining has gone in Australia, the roads and
railways have been followed by pubs, brothels and gambling
dens, as workers in isolated towns had little else to do with
their spare time and money. The vast lode of silver, lead and
zinc discovered at Broken Hill in 1884 proved to be one of
the richest ore deposits ever found in the world, and the
wealth produced by the mine led to the development of
Broken Hill—a fine city that remains to this day one of the
jewels of outback Australia. Much money earned by the miners who worked underground was spent on grog, gambling
and women. At its peak early last century, the town boasted
more than sixty pubs, half of them in the main street. Close

by were numerous discreet brothels and gambling haunts.[52] It's the same story in just about every other mining centre around the country.

In the modern 'fly in, fly out' (FIFO) era, miners can make $150,000 a year just for driving a truck, three times what a city bus driver with far greater responsibility gets paid. Many of them return to their families on their days off, after working up to fourteen days straight, and they shrewdly invest their earnings in big houses and beachside weekenders up and down the coast. But there are others who still live much like the Broken Hill miners of early last century. Some reveal details of days-long benders when they return to the big smoke, sometimes involving drugs and prostitution. David, a straight-laced thirty-something from Sydney who now works in the Queensland coalfields, says he has never met more people with broken marriages and drinking and gambling problems than during the four years he has spent in the industry. In some of the isolated mining camps west of Mackay, there's little to do at night besides drinking and playing the pokies, he says. John, who owns a farm in southern New South Wales and works as a FIFO miner in central Australia, says many of his mates fritter away their massive earnings with regular overseas trips and by going into debt to buy houses, boats and four-wheel drives. The CFMEU's Tony Maher says the industry is rife with such anecdotes, athough reliable data is yet to be produced.

In so many ways, the governments of Australia are not unlike these miners on a bender, spending the revenue from natural resources just as quickly as it comes in. But this level of spending can only be sustained if commodity prices stay high. When they come back to earth, we'll all be in for a long and painful hangover.

NICE IDEA, NEXT DECADE

It has been a long time coming, but Australians have finally begun talking about better ways to manage their squillions from the resource boom. A number of individuals and groups have called for stronger revenue management. Some have called for the creation of a sovereign wealth fund to preserve savings for future generations, while others say we need a stabilisation fund to deal with short-term fluctuations. The groups and individuals include the Reserve Bank governor Glenn Stevens, Warwick McKibbin, chief executives or chairmen of major Australian companies like Amcor, Boral, the Commonwealth Bank and Fairfax, the Australian Greens and the former Liberal leader Malcolm Turnbull. A handful of journalists have focused attention on this issue for several years. No-one has yet spelt out the hard part—how the fund should work in practice—but at least we have made a start.

With the Australian dollar super-charged by the receipts of the mining boom, it makes a lot of sense for Australia to build up a foreign currency stabilisation fund, which could over time be turned into a sovereign wealth fund for future generations. These holdings would put downward pressure on the dollar, giving other exporters a break, while also allowing us to bring some of the money back when the boom ends. If we built up a diversified foreign-currency fund worth

US$100 billion during the boom years—buying euros and yen as well as American dollars—the risk-free bonds and blue-chip equities would be worth a lot more when our currency eases. If we held onto these savings and used them when the dollar came back to its post-float average of US70c, the fund would be worth $140 billion in local currency and deliver returns of more than 40 per cent to its shareholders, the Australian citizens. If the dollar came back to US60c, as it did during the GFC, it would be worth $166 billion. Taking money out of the economy during the boom times would also take pressure off inflation and interest rates. This is what's called 'counter-cyclical' policy, saving during the boom and spending in the leaner times. The RBA already does something like this on a modest scale, buying foreign currency when our dollar is strong and then selling it to support the dollar in times of weakness. This is how the RBA generates enormous profits for the government from time to time.

With the right framework, we could create an insurance policy, a cash cow and an endowment all rolled into one. Over the long term we could accumulate savings for our children and grandchildren, as compensation for having dramatically run down the country's natural resources. The fund would also act as a second 'shock absorber' for our economy. The floating dollar is meant to be such a shock absorber, but it is now being stretched. If we had a stabilisation fund, instead of driving up the value of the currency, mining revenue would drive up the value of the fund.

This should be an absolute no-brainer, but unfortunately it is not an option our politicians seem willing to grasp. Now that the federal budget is heading back into surplus, we should start building up foreign savings while investing in projects to make the economy more productive. Even though the government is yet to pay off its debt, it makes sense to build up foreign-currency assets when the dollar is so strong. Instead our politicians prefer to spend the money under the guise of 'nation-building' and pre-election tax cuts. It is true that Australia is a growing economy and we have infrastructure bottlenecks that need to be fixed. But we could, for example, lock away a well-defined share of the windfall revenue in an offshore fund and put the rest into public works and tax cuts down the track. Our pollies, however, won't commit to such a sensible policy. Instead they want unfettered access to *our* money.

From late 2009 onwards, Glenn Stevens started prodding the federal government to safeguard our future by saving a chunk of the revenue from this boom. In one of his most important speeches on this issue, Stevens said that Australia's reliance on exports raised serious questions about how to prepare for a more volatile economy. 'The emergence of China and India is a benefit to Australia, but we stand to have a heightened exposure to anything going seriously wrong in those countries. How then to manage an income flow that is higher on average, over a long period, but potentially more volatile?' he asked.[53]

This was a salient question, which Stevens didn't answer in his speech. But when asked about it later he said, somewhat cautiously: 'Some countries take some of that income and put it offshore and find a place to put it that's inversely correlated to its source. Potentially you could think about that.'[54] This single sentence marked the first time that an official from any economic or government institution in Australia had raised the need to put windfall revenue into an offshore fund, even though Australia was in the middle of the fifth resource boom in its history. Stevens has since repeated these calls many times, along with his fellow RBA board member Warwick McKibbin, whose tenure on the board was not renewed by Treasurer Wayne Swan. McKibbin has been calling for Australia to set up a sovereign wealth fund since 2008. He says it is a simple case of putting some income 'outside the country to alleviate excess demand conditions', and by so doing helping to manage risk for the country as a whole. And he says this policy will help us to avoid becoming a 'one-sector economy'. Channelling tax revenue from a well-designed super-profits tax into an offshore wealth fund would help to achieve a more balanced economy, taking pressure off the exchange rate and off other exporters.

Former RBA deputy governor Stephen Grenville also argues that effective taxation of the excess 'rents' or profits earned by mining companies during boom times, and the creation of a foreign-currency fund, would prevent the dollar from going dangerously high. He explains:

> The excess rents are the things that are pushing up the exchange rate. If you take those out at the top of the cycle and put them into foreign currency, it would have to work. The question is how much can you extract from the mining companies—the answer so far is none.[55]

When I asked Wayne Swan to respond to Glenn Stevens's call for a stabilisation fund, he repeated his claim that super-annuation policies were the answer. 'We would see Australia save through the 8.4 million wealth funds—the superannu-ation accounts of working Australians,' he says.[56] It's a nice line, but it is dishonest and misleading. Swan was talking about his plan to increase saving in superannuation gradu-ally, mainly by increasing the compulsory levy from 9 to 12 per cent this decade, along with two other minor measures. These measures are supposed to increase super savings by a whopping $500 billion by 2030. This sounds like a lot, but in percentage terms it will add a miserable 0.4 per cent to savings each year, at a time when national income has already increased by 15 per cent. And Swan isn't even put-ting aside government revenue to fund this increase—it is coming out of the pockets of employers and, indirectly, their workers. In other words, Swan has no plan in place to deal with the windfall tax revenue he will receive from mining. Swan's 2011 budget is a prime example of the complacency that puts our economy at risk. After spending big to boost the economy during the GFC, Swan has shown no capacity to save for another rainy day now that the economy has recovered. In a budget that will spend a total of $1,558 billion over the four years to 2014–15, the net reduction in spend-ing was $5.2 billion, or a mere 0.3 per cent of the total. It is hard to imagine a weaker policy response to the structural challenges of the boom.

Joe Hockey, too, engages in some artful spin-doctoring when asked about Stevens's remarks. He acknowledges that

the idea of a special fund is appealing. 'If there's a way of
Labor-proofing a sovereign wealth fund, then you can help
to stabilise currency risk and commodity risk. That's very
attractive.' He embraces the idea of the fund as a type of
insurance policy. 'I'm a big believer in insurance. It seems
important to remove as much risk as possible that can affect
exporters.' But setting up such a fund is 'the Maserati of
public policy', he goes on—everyone wants one but no-one
can afford it. While he has a great deal of respect for Stevens,
Hockey says, his comments were about a 'far, far away land'.
There are too many other priorities, like investing in infrastructure and delivering another round of tax cuts. A stabilisation fund might be a good idea, but it would have to wait
for another decade, when all Commonwealth debt has been
repaid. 'I look forward to it in my tenth year as Treasurer,'
he concludes.[57] Hockey's comments are misleading because
he is focusing on gross Commonwealth debt, ignoring the
debt reduction that would arise from accumulating foreign-
currency assets while the dollar is strong, thereby reducing
net debt.

The Opposition leader, Tony Abbott, also dismisses the
idea as premature, given that the federal government could
be paying off the GFC debt for another decade. He even
goes so far as to describe it as a 'distraction'. 'I think it is a bit
premature to be talking about sovereign wealth funds,' he
says. Abbott holds up the Coalition's role in creating the
$75 billion Future Fund as an example of its commitment to
such measures, although this fund is only designed at present to offset the cost of public-service pensions. Asked to

explain why the Howard government did not do more to lock away savings during its last years in office, Abbott says people would have been 'screaming' for tax cuts had the government been running surpluses of $40 billion a year.[58] Or perhaps it was the Liberal Party strategists who were screaming for the government to spend the money. A few weeks after giving this interview, Abbott began flagging a new round of pre-election tax cuts.

The leading conservative thinker Phillip Blond, who has been a driving force behind the 'big society' agenda in the UK, urged his conservative cousins in Australia to dump the welfare state and adopt a Norwegian-style savings policy during a June 2011 visit. He told Liberal frontbenchers that Australia had blown its last two resource booms and that without such a policy was on track to blow this one. But Blond's good counsel was met with blank stares. 'Well, I mean, everyone nods, says "Yes, it is true," and moves on. So it is one of those interesting things that everyone acknowledges should be done but no-one is leading with it,' he says.[59]

As a last resort, I approached one of Labor's bright young sparks, Andrew Leigh, an economics professor who has a Harvard PhD and is now a federal MP, in the hope of getting some sensible comment. Leigh admitted that there was a compelling case for investing in foreign-currency assets while the dollar was so strong, but he then rolled out the party line about investing in infrastructure instead, as this would deliver 'higher rates of return'. I should have known what he was going to say when he gave me his business card: it has a picture of him wearing a hard hat and an orange fluoro vest. More recently, Leigh has ridiculed the idea of a sovereign wealth fund in an opinion piece based on the dubious assumption that 'future generations will be richer than us'. He argued: 'So there is no philosophical obligation to leave our children an overstuffed piggybank rather than a good education and a well-functioning rail network.' Indeed, Leigh even claimed that 'future generations might well condemn us as short-sighted scrooges'.[60] One big problem with Leigh's argument is that we don't know what rate of return we are getting from infrastructure investment. As Future Fund chairman David Murray says: 'One thing we don't have is an accepted model for determining the cost-benefit of infrastructure. There are models out there. We haven't hit on one.'[61] What's more, Leigh's argument ignores the inflationary effects of spending *all* the windfall revenue. His plan to invest in a rail system means we won't have savings that can be drawn on to pay for stimulus in the next downturn, or to pay for fixing the damage wrought by climate change.

The bottom line is that neither side of politics has an interest in implementing sensible policies. There is nothing to prevent a repeat of the Howard government's $300 billion spending orgy. The sad truth is that despite the efforts of the RBA governor and others, this debate is unlikely to take us anywhere until we change our political culture.

YES, TREASURER

The situation is complicated by the fact that the RBA's responsibility is limited to monetary policy, while its Canberra-based rival, the Treasury, is the nation's premier institution when it comes to economic policy. If Treasury gets it wrong, the RBA will be unimpressed and we will pay higher interest rates as a result. As it turns out, Treasury got it wrong last decade with its advice on the China-driven boom, and it looks as though it is getting it wrong again. It is worth revisiting Treasury's role in the Howard government's decision to open the fiscal floodgates at a time when the economy was booming.

When the mining boom Mark I got underway in the middle of last decade, very few of our economic experts and none of our politicians discussed how Australia should manage its luck for long-term gain. The prevailing view was that this sudden burst of prosperity was destined to last forever. In the lucky country, complacency ruled.

When Prime Minister Kevin Rudd held his '2020 Summit' in 2007, not one of the thousand or so luminaries drawn from across the country raised the issue of how better to manage our commodity income—even though the mining boom was in its fourth year and was looming even larger. The economic discussion, chaired by former Treasury economist and Westpac chief David Morgan and Treasurer Wayne Swan, did not even mention the mining boom, and nowhere in the 400-page report is it acknowledged. It was as though everyone assumed the boom would continue indefinitely.

Professor Ross Garnaut was one leading figure who became alarmed by the extent to which money was being pumped into tax cuts and middle-class welfare, when common sense indicated that much of it should be saved. As we have seen, Garnaut is well versed in the potential pitfalls of resource wealth, as he worked in Papua New Guinea before and after independence and helped to establish that country's stabilisation fund. In 2005 he delivered a paper to a conference attended by many of our business leaders and top economic experts. In this paper, called 'Breaking the Australian Great Complacency in the Early Twenty-first Century', he urged the government to stabilise the economy by saving the entire windfall increase in tax revenue from the mining boom. This revenue, Garnaut argued, was then approaching 2 percentage points of GDP—or about $25 billion per year—and should all be saved. 'Prudence argues for this windfall to be saved in fiscal surpluses. This makes the recent surpluses in public budgets seem far too low in the context of the stabilisation requirements, at a time of booming domestic demand,' he said. Even though the conference was attended by some of the top experts in Treasury, Garnaut's warning was not taken seriously. After the speech a senior Treasury official told him the department could not do any better because the government was not prepared to produce higher surpluses. When he outlined similar arguments at an in-house Treasury seminar in 2005, one official responded to Garnaut's call for massive surpluses by saying, 'You are trying to turn us into Papua New Guinea.'[62]

Garnaut now believes Treasury was complicit in encouraging this spending frenzy. Looking back on these events, he says that Treasury 'provided arguments for spending rather than saving'. The justification for the extra spending reflected a serious flaw in Treasury's thinking. The department argued in budget papers at the time that the floating exchange rate would insulate the economy from the huge swings in national income. Garnaut summarises the thinking as: 'You spend it, the exchange rate goes up. The floating exchange rate stops this being inflationary.'

The department was caught red-handed providing such excuses when officials wrote in advice in June 2007 that the resource boom could be considered 'permanent'. In a minute to the treasurer of the day, Peter Costello, Treasury wrote: 'There is a strong case for spending additional revenue from the increased terms of trade to the extent that the increase is considered permanent.'[63] With such advice, it is little wonder that Costello presided over the biggest spending binge since the Whitlam years.

After five years of strong demand for minerals, Treasury argued in a 2008 paper—its first piece of considered research on the resource boom—that Australia should not fear the 'resource curse' and could sit back and enjoy higher living standards. The paper was written by middle-level officers, but it was reviewed by some of the department's most senior staff. The paper argued that this boom might last a very long time, and it included some bold claims about rising affluence from mineral wealth:

> The prospect of the rise in the terms of trade being sustained therefore need not be considered a 'resource curse' that will simply create problems. If well managed, the transition to higher terms of trade presents an opportunity to raise Australian living standards.[64]

Note that the paper was released just three months before the global financial system suffered its biggest collapse since 1929. The release of this paper was definitely a case of bad luck—or being duped by too much luck. Subsequently, our economic experts in Treasury have been even more optimistic about our future, declaring that the boom could span 'several decades'.[65]

Garnaut is less euphoric. He fears that Australians are in for some very tough medicine when commodity prices eventually come off the boil, as current levels of expenditure are 'only sustainable in very good times'. Garnaut is highly critical of Treasury's role in last decade's boom. He says: 'Treasury got it wrong. They believed that the modest surpluses we had weren't too bad. On the question of the need for much bigger surpluses there was a strong push back from Treasury on that. They didn't have their heart in it.' Even more alarming is that given the projected revenues from the mining boom this decade and next, and the lack of commitment from our politicians to serious saving, such challenges are likely to recur.

LEARNING FROM BITTER EXPERIENCE

Australia's failure to generate better policies for its resource income is particularly poor form given the examples of best practice available around the world. Like many Latin American countries, copper-rich Chile has been through some very tough times since the early 1970s, not only because the US-backed President Augusto Pinochet's reign of terror led to the execution, torture and disappearance of thousands of suspected opponents. The collapse in copper prices from the late 1970s onwards meant the country had to contend with the Washington-based IMF as well as the CIA. In the early 1980s, when Australia suffered its worst recession in fifty years, Chile endured a tumultuous collapse in copper prices that, when combined with interest-rate rises, led to a severe contraction in the order of 12 per cent of GDP, or about six times the contraction experienced by Australia at the same time.[66] Like Zambia and other resource-rich economies, Chile suffered from a false sense of optimism. Australian policy-makers should take note, for we appear to be showing similar signs of smugness. 'It is ironic but salutary that the optimism engendered by the expected post-1978 mineral boom prompted self-defeating policies,' notes Professor Richard Auty, who has examined the experiences of many such countries.[67]

But now, Chile has emerged as a standout example of how to ride the resources rollercoaster successfully. Australia and Chile are remarkably similar economies. While Chile's mining sector is more narrowly focused on copper, its share of exports, and of GDP, is almost identical to Australia's. The big difference is that Chile now saves a share of its mineral income in a disciplined way. In 1985 Chile set up its Copper Stabilization Fund to bank windfall earnings and by the middle of the last decade had accumulated US$5 billion in savings. From 2006 the government of President Michelle Bachelet put in place even tighter savings plans, including the 2006 Fiscal Responsibility Law, which involved the creation of two new sovereign wealth funds. The first, the Pension Reserve Fund, is quarantined for at least ten years. The government jump-started this fund with an injection of US$600 million, and it is topped up each year with tax revenue worth between 0.2 and 0.5 per cent of GDP, depending on the size of the budget surplus. The second fund, the Economic and Social Stabilization Fund (ESSF), replaced the original Copper Stabilization Fund. It receives money when fiscal surpluses rise above 1 per cent of GDP. The funds adopt a low-risk invest model, with about two-thirds held in risk-free bonds and 30 per cent in money-market securities such as bank bills. Significantly, all of the assets are held in foreign currency, with about 50 per cent held in US dollars, 40 per cent in euros and 10 per cent in Japanese yen.

With these controls in place, Chile was able to build up significant savings during the boom last decade. The country succeeded in quadrupling the size of its assets to more than $20 billion, worth more than 10 per cent of national income, thus creating a significant buffer against a global downturn. Had Australia saved the same relative share of income we would now have a fund worth $130 billion.

Andrés Velasco, who was finance minister of Chile from 2006 to 2010, explained the thinking behind these funds. As finance minister, he told Chileans after the GFC that the government would run a disciplined counter-cyclical policy, saving aggressively during the boom years and spending during downturns. Velasco, now a professor at Harvard University, argues that the policy 'resulted in steady gains in quality of life and benefits that do not depend on the short-term situation in the global economy or change with swings in export prices'. The funds enabled the country to weather the GFC without going into debt. Even more impressive than the design of the fund was the government's success in building consensus for the policy. Velasco points out that the debate has reached a point whereby 'what once seemed controversial is now established wisdom'.[68]

At exactly the same time as Chile was formulating and implementing this bold policy, John Howard's cabinet was drawing up plans to unleash Whitlamesque spending. The government was fanning the flames of inflation, forcing the RBA to put up interest rates. Peter Costello has tried to claim that he acted responsibly by setting up the Future Fund. With the right policies this fund could be turned into a stabilisation fund to help manage the economic cycle and save for future generations—but Costello put no such arrangements in place.[69]

Asked to respond to the proposition that the Howard government blew the money from the last decade's boom, Future Fund chairman David Murray says: 'That's a reasonable argument. I thought when the Future Fund started it should have had $200–300 billion in it ... much higher budget surpluses going in there.' Murray argues that 'Australia Inc.' is particularly at risk because we have variable revenue from resources and high fixed costs from our welfare state. This reinforces the need to save, he says:

> With resource extraction and volatile terms of trade there is a budget-stabilisation issue to address. Secondly,

with resources and the risk of Dutch disease in the economy there is both an inflation-management issue and an intergenerational-wealth issue. There are many countries in the world with small populations and large resource endowments who have confronted that problem by setting significant sovereign wealth funds. There are some other countries with surplus reserves that set them up on a rainy-day basis. In Australia there is an argument for stabilisation and a strong argument for intergenerational wealth building.[70]

Oil-rich Norway provides another example of a better approach. Norway and Australia have more in common than most people think. The two countries vie for first place in the United Nations' comprehensive measure of quality of life. The UN Human Development Index tallies the spectrum of measurable factors that underpin human existence. The annual study has in recent years found that a combination of resource wealth and first-world institutions has proved to be the sweet spot, elevating quality of life in the two countries to the maximum obtainable. But Norway differs from Australia in one very important way—it is certain to remain in its enviable position long after its resources have been depleted. Australia has no such certainty.

Norway has a lot of oil, but that's not all. It has also had great leadership, including that of former prime minister Gro Harlem Brundtland, who served as a Labour Party prime minister three times between 1981 and 1996. During this period, the government and parliament of Norway developed policies to safeguard its resource revenue. More than anyone else, Brundtland articulated the concepts of sustainable development and inter-generational equity— the idea that the current generation should not leave future generations worse off. In the early 1980s she set up the World Commission on Environment and Development, which is often referred to as the Brundtland Commission. In 1987 the commission published the report *Our Common Future*, which spelled out what Bruntlandt meant by the concept of sustainable development.

Brundtland and others in Norway had seen the damage done to other countries by mismanaged resources income. Taking on board the principles of inter-generational equity, they set up a petroleum fund that is as breathtaking as it is simple in design and purpose. In essence, Norway's fund transforms a non-renewable resource into a financial asset that can last forever. It does this through the principle that no more than the 4 per cent of the fund should be spent in any one year.

Norway discovered oil in the North Sea in 1969 and the country soon realised that it was going to be inundated with money. As the former central-bank governor Svein Gjedrem put it, the country realised that 'the revenue would transform Norwegian society'. In 1983 the Committee on the Future of Petroleum Activity proposed a stabilisation fund, but towards the end of the 1980s this thinking evolved into a fund designed to hand wealth to future generations. The conservative government passed the Act on the Government Petroleum Fund in 1990, although as a result of the global recession the country was then in deficit. It wasn't until 1996, Bruntland's last year as prime minister, that transfers into the fund began in earnest.

Just as Eskimos living in the Arctic store food for the long winter, Norway has been stashing the cash from its oil riches to assure its future. Tiny Norway—a country of just 4.8 million people—now has the second largest sovereign wealth fund in the world. Since 1996 it has accumulated more than US$585 billion in savings, even more than Saudi Arabia's oil fund. At the present rate of saving Norway's fund will double by 2020, and it will most likely keep doubling every seven to ten years as compound interest kicks in. The country also manages to save a lot because it maximises its tax take from local and foreign oil companies. Its corporate tax rate is 28 per cent, but oil companies pay an additional 50 per cent tax for the privilege of profiting from the wealth of Norwegians.

Norway's fund has already proved enormously beneficial to present-day Norwegians, even though the main aim is to hand wealth to future generations. During the fifteen years that it took Norway to build up assets worth almost $600 billion, Australia has accumulated an additional $485 billion in net foreign debt, lifting the total to $677 billion in March 2011. A study by the National Institute for Economic and Industry Research found that Norway had consistently generated trade surpluses since the 1980s, whereas Australia had accumulated deficits. Norway's management of its natural resources had generated US$150,000 per person more in international financial assets than Australia had managed to accumulate.[71] Norway's fund also helps the country to maintain a diversified export base. All of its assets are invested in foreign currency, thereby taking considerable heat out of the krone. Despite its significant oil revenue, the resource sector only accounts for half of the country's export earnings. Norwegian exports are still affordable to other countries and include world-renowned products like smoked salmon and cheese, but also ships, pulp and paper products, metals, chemicals, timber and textiles.

After initially investing in risk-free government bonds, in 1998 Norway's fund began investing in equities and in 2002 non-government bonds. It now has 60 per cent of its assets in global equity markets, about 35 per cent in bonds and 5 per cent in real estate, which was added to the mix in 2008. The equities and bonds are spread across a range of markets and currencies, thereby reducing risk. The policy also gives Norway enormous financial power and flexibility. After the GFC, the fund snapped up blue-chip stocks, lifting its annual return to 25 per cent in recent years. It also has power to influence corporate decisions by dumping the stocks of companies that engage in unethical behaviour. Since introducing an ethical code in 2004, the fund has sold its shares in more than twenty-five companies that had been involved in human rights abuses, environmental damage or the production of armaments. These companies include some of the biggest corporate entities in the world, including businesses operating in Australia. Lockheed Martin, Raytheon and EADS were dumped because they made cluster bombs or nuclear weapons. Wal-Mart was dumped for human rights abuses and Rio Tinto for causing environmental damage in Indonesia. Serco, which runs detention centres in Australia, and Thales, a major defence company, are also on Norway's ethical blacklist.

Norway and Chile have both found ways to avoid the common dangers of the resource curse. One of the weaknesses of both models is that such funds could conceivably be raided by future governments. Their only safeguards are transparency and an extraordinary degree of public consensus. The US state of Alaska has gone even further, introducing a constitutional amendment to establish its Permanent Oil Fund in 1976. The amendment requires that 25 per cent of all mineral revenue is placed into the fund. The principal must be preserved, and the fund pays out dividends to each citizen based on annual earnings. This policy came about after public concerns that the government had wasted the money it gained from the leasing of oil fields in the late 1960s. In 2010, the dividend paid was $1,281 to every woman, man and child, including 'resident aliens and aliens'. One of the most impressive things about this fund is that income from oil taxation, and from investments, will continue to rise even as oil production declines. Such are the benefits of making an early start and setting up a savings plan long before the resources begin to decline.

AND NOT LEARNING

With such impressive international experience to draw on, it is surprising that Australian experts, politicians and policy-makers haven't thought to apply such lessons here. Yet federal parliament's economics committee has never been asked to examine these models, nor have our politicians visited places like Norway, Chile or Alaska to learn from their experiences. Every winter, federal politicians jet off overseas on lengthy jaunts which are ostensibly about bringing back new ideas to Australia. So far, none of these tours has been aimed at learning how better to manage the mining boom or the money derived from our resource wealth.

Ross Garnaut, who has been immersed in the economics of resources for forty years, has not previously advocated a special fund to save windfall revenue, but he thinks Australia is now crying out for one. 'This boom has been so big it would have been better for us to save. That would have stopped the real exchange rate going so high, and it wouldn't have flattened the traded industries outside mining as much as it has,' he says.

Garnaut fears that Australia lacks the capacity to manage its wealth, and that we are headed for a painfully hard landing. 'We are in danger of dissipating the benefits of this boom, and then the peak will pass and we will find it difficult to adjust down.' Asked why solutions like a stabilisation fund are not being advanced, he says, 'that is all part of the complacency'.

Around the time Garnaut was raising his concerns about mismanagement, I returned to Australia after several years overseas, including three years working in East Timor. There was a dearth of debate about what to do with the billions rolling into the Treasury coffers from the China boom.

Newspaper columnists clamoured for tax cuts and infra-structure; the debate was all about how we should spend the money. Having been immersed in the planning of East Timor's oil fund, I found the lack of any debate about long-term sustainable management surprising. In late 2006 I wrote the first of many opinion pieces on why Australia should look at these experiences and design policies to manage its increased revenue. I decided to find out what our economic experts thought of this idea by using the Free-dom of Information Act. The result was a set of twenty-nine Treasury documents released in 2008, which showed that, surprisingly, some officials had paid considerable attention to the overseas experience. One research note by Kirsty Laurie, the Treasury economist who also happened to have co-authored the paper on Howard's spending binge, said that investing all the oil revenue in a foreign-currency fund 'pro-motes exchange-rate stability'. Laurie concluded: 'Norway appears to be very conscious of avoiding the Dutch disease so that restructuring costs are not excessive when petroleum revenue declines.' Another Treasury paper, 'Commodity Hedging for Government' by Michael Bath, said that when countries are inundated with windfall revenue they 'strug-gle to avoid pro-cyclical fiscal policies', which generally leads to higher inflation and interest rates. Sound familiar? Bath concluded that the aims of the Norwegian fund were indeed profound. 'It transforms the wealth associated with a non-renewable, finite stock of resources into a perpetual source of relatively steady revenue,' he wrote, before concluding that Norway provided 'an excellent case study in managing commodity risk in a manner that maximises fiscal sustainability'. Other documents released by Treasury discussed the benefits of PNG's approach.

These papers seem to indicate that Australians are getting value for money from their Treasury economists, although the department also showed a curious bias against sovereign wealth funds. A November 2007 briefing note for officials attending the Group of 20 meeting reinforced US fears about the power of such funds, which now have a combined value of more than US$4 trillion. At the time, the then Federal Reserve chairman, Alan Greenspan, had expressed alarm about the undue power of these funds, and Treasury echoed these concerns. The note observed that there were 'concerns due to the sheer size of these investment vehicles, their lack of transparency, their potential to disrupt financial markets, and the risk that political objectives might influence their management'. Australia has since taken an active role in global efforts to improve the governance of sovereign wealth funds. David Murray, chairman of the Future Fund, also chairs the International Forum of Sovereign Wealth Funds.

But it seems that none of Treasury's analysis was influencing the policy process, as the notes written by Laurie and Bath never actually went to the treasurer as policy briefs. More recently I asked Treasury again what advice it had been giving the government on this issue. It turned out that there was very little. One note written for a minister attending a conference on sovereign wealth funds in May 2010 advised that he might get asked whether Australia should follow Norway's example. He was told to say: 'The first observation

I would make is that Norway's circumstances are different from Australia's. Norway has a single resource in its interest in North Sea oil. Australia has more diversified and longer-lived natural resources.'

In the 2011 budget, Treasury discussed these issues in a special statement about what it called the challenges and opportunities of 'an economy in transition'. In a lengthy discussion the department dismissed the threat of Dutch disease in 'advanced countries'. To bolster its argument, Treasury cited Norway as an example of one Western country that had successfully avoided the resource curse. Norway had maintained 'a well functioning non-oil traded goods sector' and its manufacturing had benefited from the impact of higher oil revenues. But this was very selective analysis on Treasury's part, for it failed even to mention in passing the fundamental role played by the country's $600 billion oil fund to defend against the ill-effects of resource wealth.

Just as Treasury found excuses to justify the Howard government's reckless spending, it is now finding excuses for the government to avoid taking prudent action to safeguard our future.

Tax Us if You Can

'Policy-makers around the world can learn a lesson when considering a new tax to plug a revenue gap, or play to local politics.'—Rio Tinto CEO Tom Albanese, July 2010, one week after Labor dumped Prime Minister Rudd and the super-profits tax[72]

In mid-2010, when the world's biggest mining companies were experiencing their best trading conditions ever, they suddenly ran into trouble. While voracious demand from China and the rest of Asia had driven prices and volumes to record levels, the big miners were alarmed by a plan unveiled by Australia's Labor government to impose a 40 per cent 'super-profits' tax on their earnings. Not only would they pay more tax in Australia; they also feared other countries would follow suit. Before they knew it, an outbreak of resource nationalism could hit their global operations.

Mining companies usually benefit from the belief that what's good for the company is good for the country, especially in the developing world, where they exercise a lot of influence over governments. But in this case they were not dealing with your typical tin-pot resource-rich country. They were dealing with Australia, a $1.3 trillion economy and one of the world's biggest mineral exporters, a country that had become the only Western nation to sail through the global financial crisis and record its nineteenth year of continuous economic growth. The amount of money at stake in Australia alone was mind-boggling. The miners had put in place expansion plans to at least double coal and iron-ore production in coming years. In 2010 these two commodities were already Australia's most lucrative exports, earning around $90 billion a year—three times the amount earned a decade earlier. Combined with soaring prices, this expansion would deliver astronomical returns on the miners' operations.

As the mining boom gathered unprecedented momentum, the then Treasury secretary, Ken Henry, realised it was time to apply a more rational and effective tax regime to resource production. As a boy growing up amid the tall timbers of Taree in northern New South Wales, Henry had seen how logging companies were able to harvest century-old trees in exchange for a royalty of a few dollars.[73] He realised that resource companies in Australia were doing pretty much the same with assets that, unlike trees, cannot be replanted. Economists in the federal Treasury had estimated that during the resource boom in the 2000s, state and federal governments had kissed goodbye about $35 billion in revenue because they failed to tax the super-normal profits that arise from time to time. This is why they came up with the so-called Aretha Franklin tax—the resource super-profits tax, or ReSPecT. Treasury put together a proposal that was far more complex than it needed to be, given they had a workable model in the petroleum resource rent tax

(PRRT), which was introduced back in the 1980s. Their new proposal involved the government taking a 40 per cent stake in resource projects, sharing in both the losses and the profits, which opened up a Pandora's Box. It was a good idea that was poorly designed and communicated.

In his first interview on these events, Kevin Rudd reveals that the proposed tax followed 'long-standing research into the problem of Dutch disease' by Treasury, and that Henry put forward the proposal in this context. 'I knew what he was talking about, the dual effect of putting capital into a globally competitive sector, leading to the massive appreciation of the currency, leading to a double impact on the rest of the economy as it adjusts to the exchange rate while starving it of investment capital,' Rudd says.[74]

The new tax would have raised \$12 billion in the first two years, rising to more than \$100 billion over a decade. But it wasn't designed only to generate money; it also aimed to stop our economy from, so to speak, losing its head, by ensuring a more balanced development of our non-renewable resources. As soon as Rudd sprang the new tax on the industry, the big three companies decided they had to kill this plan—and they were prepared to play dirty. When London-based Rio Tinto, Melbourne-based and London-listed BHP Billiton and Swiss-based Xstrata put their collective weight together, they are a formidable combination. Their total combined value on global sharemarkets is \$450 billion, 86 per cent of which is in foreign hands. The three companies are worth more than the size of Australia's federal budget, and about one-third the size of the entire Australian economy.

Together they embarked on a savage lobbying effort to bring down the proposed tax by attacking the government and its prime minister. They began this extraordinary campaign before the proposal had even been put into legislation, and before parliament had had the opportunity to review it.

BHP led the offensive, establishing a 'war room' inside its Melbourne head office. Run by senior financial executive Gerard Bond, along with senior staffers and external consultants, this team worked on the project for about seven weeks. BHP commissioned its own focus-group research, which was used to drive a $22 million TV and print-media blitz and a targeted lobbying campaign that included Geoff Walsh, a former national secretary of the ALP and former staffer to prime ministers Bob Hawke and Paul Keating. BHP spared no expense on the campaign, which reported directly to CEO Marius Kloppers. External talent included the market-research specialist Tony Mitchelmore and the corporate strategist John Connolly. Mitchelmore had been plucked from obscurity by Labor to work on the Kevin07 campaign and had stayed on doing qualitative research for the ALP before working for BHP on this campaign. He organised an intensive round of sixteen focus-group sessions, which revealed that many participants believed Rudd's proposal had come from left field and was likely to derail the one industry that was keeping Australia's head above water. Realising that they had a good chance of killing the tax, the miners adopted a 'whatever it takes' approach. Mitchelmore conducted further focus-group research for the next six weeks, and the findings informed the TV advertising campaign crafted by another Kevin07 veteran, Neil Lawrence.

The miners' efforts were spectacularly successful. Seven weeks and four days after unveiling the preliminary plan, Prime Minister Kevin Rudd was deposed and so was his tax. Despite having guided his country through the worst global downturn since the Great Depression, he became only the second first-term Labor prime minister to be removed from office, the first being James Scullin in 1930. Big Dirt, as the three companies are now known, executed regime change two months before the voters of Australia exercised their democratic rights at the ballot box. Having subverted a functioning democracy, mining executives were celebrating in airport lounges around the country.

Rudd's plan for a 40 per cent marginal tax on super profits was not some kind of left-wing tax grab. Some of the world's more conservative economic institutions, such as the IMF and the World Bank, say it is good policy. While the version proposed by Rudd suffered from poor consultation and complex design, it was still at the drawing-board stage. The government wanted to overhaul a mishmash of unwieldy state royalties that harked back to the nineteenth century. In its place, they would introduce a modern and more rational system, to be run by the federal government. Miners would be compensated for the payment of state royalties, but the overall tax burden would be higher, around 57 per cent. The new tax would also have delivered a substantial windfall to the current generation and, if the government managed the money well, to generations to come.

Immediately after becoming prime minister on 24 June, Julia Gillard turned her attention to thrashing out a deal with the three multinational miners. Eight days later, she announced a breakthrough that cut the marginal tax rate from 40 to 22.5 per cent, restricted its scope to coal and iron ore, and added some creative accounting concessions for the big miners. Total cost of the concession: $15 billion over four years, rising to $60 billion over ten years (and possibly $100 billion if prices stay high). Gillard changed the name of her policy to the more polite-sounding Mineral Resource Rent Tax (MRRT). A raft of emails released under FOI shows that BHP was very much running the show. Its executives drafted the heads of agreement before emailing it to Wayne Swan's office for approval.

Repeating her 'moving forward' mantra, Gillard announced the compromise like this: 'It moves things forward whether you're a coal miner in the Bowen Basin, a contractor in Karratha, an opal miner in Coober Pedy or a young worker in Sydney.' In fact, the MRRT deal made life worse for smaller Australian-based miners by removing the resource exploration rebate and by awarding big miners a significantly lower tax rate. For iron-ore miners with mature projects, which means the big companies, their projects would be taxed at 36.4 per cent—close to or even below current levels—whereas small or medium-sized projects would pay an average rate of 48.9 per cent, according to modelling produced by Treasury and released under FOI. The big miners benefited from a concession that allows them to calculate deductions for tax purposes using the market value rather than the purchase price (or 'book value') of their assets, providing huge depreciation allowances. The small and medium Australian players were not represented in the negotiating room, and the new deal actually reversed the central and laudable aim of the RSPT—that is, reducing the tax burden on start-up operations, which are penalised by the state royalties because the impost is paid when production starts, rather than after the company actually begins to make a profit. The success of multinational miners in securing these concessions, and in beating voters to the punch, reveals the perverse world order in which we live: an advanced country can possess enormous riches but lack the capacity to do what is clearly in its own long-term interest. Resource-rich countries often turn out to be very poor at managing their abundant wealth, and this doesn't apply only to the developing world. As this case demonstrated, muscular multinationals can be very effective when bearing down on weak Westminster governments.

Not only did the miners change the prime minister and change government policy, they went on to brag about how their coup had stopped similar schemes from spreading around the world. When Australia moved to introduce the RSPT, African and Latin American countries were paying close attention. Kevin Rudd says the vicious campaign by the industry established a precedent. The defeat had serious implications in Africa, where mining companies trade countries against each other, and in Latin America. Rudd recalls:

The stakes were phenomenally high. Which is why they wanted any significant tax adjustment stopped dead in its tracks here, despite their record profitability. They feared the precedent it would establish in the developing world, where decent revenue flows are needed desperately to fund the building of the most basic infrastructure to underpin their long-term economic development.[75]

Exactly one week after Gillard announced the compromise, Rio Tinto's American chief executive, Tom Albanese, told a group of mining executives in London that the Australian experience should send a salutary message to governments around the world. Governments should 'learn a lesson' from the episode, he declared. A few months later, Xstrata's chief executive, Peter Freyberg, was still bragging. Xstatra is partly owned by the shadowy Glencore International, a commodities trading house that was founded by Marc Rich, who fled to Switzerland after being charged by the US government with tax evasion. Glencore's modus operandi is to profit from control of the supply of key commodities, and it has been accused of creating speculative bubbles. Interviewed on national radio, Freyberg stressed that his company could move investment offshore if local policy didn't suit. Asked if he was threatening Australia, he said, 'No, it is not a threat. It is a statement of fact ... We have investment opportunities globally for be it copper or coal or other commodities, and we look at investing in the most competitive regions we can to maximise returns for our shareholders. If policy makes Australia less competitive then we will see that investment move elsewhere.'[76] This raises one of the real furphies in the mining companies' propaganda. Mines are not like factories that can be packed up and moved offshore. The resources are here in the ground, and Australia provides a stable, peaceful and democratic environment. Places like Brazil and Africa provide potentially more supply, but companies have to factor in higher 'country risk' in many of these places. Australia is the sweet spot for mining multinationals: we have a lot of the stuff and we let it go for a song.

BHP's executives managed to avoid bragging, although this company did more than any other to bring down the tax and Kevin Rudd. The total cost of the campaign was $22 million. The Minerals Council of Australia, which is largely funded by the big three companies, spent $17.2 million, while BHP spent $4.2 million on its own and Rio $537,000. Cabinet ministers in the Gillard government say that Geoff Walsh delivered the Mitchelmore research directly to the then ALP national secretary, Karl Bitar. These claims are strenuously denied by Walsh.[77] But the BHP research is understood to have panicked the Labor heavyweights, prompting them to move against Rudd even though he still had a commanding 4 percentage point lead in the national Newspoll.[78]

The campaign by the mining companies was aided by the federal Opposition and the nation's peak business group, the Business Council of Australia, even though the non-mining sector would have gained a 2 percentage point cut in the corporate tax rate from the proceeds of the new tax. The Business Council is meant to represent the full spectrum of business in Australia, yet its position was determined by an industry representing just one-tenth of the economy. The bankers who dominate the BCA, most notably president Graham Bradley, no doubt thought they'd see fewer financing deals as a result of RSPT. The Opposition, meanwhile, was wholeheartedly with the miners. During the fracas, Tony Abbott gave the industry his undying support, declaring he would fight the tax as long 'as there is breath in my political body'. Abbott's party has plenty of form when it comes to taking care of miners. In 1991 it opposed an entirely reasonable and long overdue policy to end the tax exemption on gold mining, while accepting donations from gold miners. The support of the Coalition and business for the mining industry shows what can happen to good policy when powerful interests successfully divide and conquer the political system.

This episode highlights the growing power and influence of mining and energy companies. Increasingly, these companies control our economy and environment in ways we have scarcely begun to comprehend. Mining's rapid growth has given it a much bigger say in national affairs and key figures like iron-ore billionaire Gina Rinehart want even greater influence, which is why she has taken a stake in the Ten Network and Fairfax Media. No longer do miners want to be seen as just diggers of dirt. They want to be recognised as corporate titans in charge of highly sophisticated operations. Western Australian premier Colin Barnett has expressed this view, demanding more respect for the industry: 'I was somewhat offended when Western Australia was described as China's quarry, and that sells this state short. The industry here is world leading, very sophisticated, high technology and the mining and petroleum industry is Australia's leading industry. This is the future income for all Australians,' Barnett said on national television.[79]

A TAXING INVENTION

The super-tax episode is not an isolated case. Australia has a long history of failing to exercise strong governance over mineral and energy resources. For over 150 years this country has celebrated those who are able to hold government to ransom and refuse to pay for the right to exploit our wealth, starting with the miners at the Eureka Stockade, who revolted over a plan to impose higher taxes. In the latest episode, billionaire miners stood shoulder to shoulder with workers in hard hats, as though they had a common interest.

One of the legacies of Australia's 1901 Constitution is that state and territory governments control the development of mineral concessions. These governments have proved particularly inept at taxing the industry. They say you should never stand between a state premier and a bucket of money, but when it comes to mining, it's a special case. Taxes and royalties haven't kept up with the huge profits now being earned. For example, when revenue earned by mining companies increased three-fold over the course of the last decade, the royalties paid to state governments increased only 2.3 times.

It is no exaggeration to say that every state government in Australia has been letting down its citizens, especially since the 1970s, when a better way to tax mining emerged. From 1973, Australian economists Ross Garnaut and Anthony Clunies-Ross set out in economic literature how a resource rent tax (RRT) would capture a fair share of profits without undermining investment in the industry. In using the term 'rent', they were targeting the very high profits that accrue when companies are given control of a valuable commodity of which there is finite supply. The two had been working in Papua New Guinea, where the Bougainville mine had been given extended tax exemptions by the colonial administration. As Garnaut explained, this mine, which began production in 1972, soon benefited from very high copper and gold prices. The site was so profitable that half of Rio's profits came from this single mine, although the company only owned half of it. 'They'd been given a tax-free holiday, been able to hold depreciation, and then gold was going to be tax-free. It was an extraordinary agreement,' he says.[80]

The two men set their minds to devising a way of collecting a fair share of Bougainville's profits without discouraging investment in other mines. They came up with the RRT, a tax that only kicked in when returns exceeded the threshold needed to justify the investment in the first place (they put this at 10 percentage points above the long-term risk-free bond rate). The art of taxation has been described as plucking the goose in such a way as to maximise the quantity of feathers obtained and minimise the hissing. Garnaut and Clunies-Ross had devised a no-hiss tax, or so they thought. The federal Labor Party adopted the principle of the RRT as part of its platform, although the state branches were evidently not paying attention. Not long after winning office in 1983, the Hawke government set about introducing the RRT for the offshore petroleum industry, which was the only area where the federal government had jurisdiction over mineral resources. The proposal did in fact create a lot of hissing, with the big oil companies declaring that the tax would kill the golden goose. The shadow Treasurer, John Howard, warned that 'the Hawke government's RRT will effectively destroy the incentive for offshore exploration'.[81] And Alexander Downer, the newly minted member for Mayo, gave Labor an absolute roasting in parliament:

> I think it is an extremely regrettable proposal; I think it is an ill-considered proposal; and, what is worse, I think it is an ideological proposal which is going to do very real damage to oil exploration in Australia … It is an anti-production, anti-development, anti-profit and anti-export tax, and the government deserves to be condemned for the irrationality of that decision.[82]

There was nothing irrational about this tax at all; it was rational economics at its very best. Bob Hawke, together with his treasurer, Paul Keating, and his resources and later finance minister, Peter Walsh, persisted—Labor politicians had more backbone in those days, as well as some good advisers who persuaded them to stay the course. Eventually, the oil companies realised that the new impost, which became known as the Petroleum Resource Rent Tax (PRRT) and was an adapted version of the Garnaut model, was well designed. It was so well designed that in 1990 BHP and Exxon asked the government to replace production royalties on their Bass Strait operations with PRRT, thereby enabling them to extend the life of those fields by twenty years or more. Trade Minister Craig Emerson, who wrote his PhD thesis on RRT, said the regime has 'stood the test of time, having scarcely been modified in its twenty-five years of operation'.[83] It did not stop some very substantial investments in offshore gas production, including $60 billion in the recent Gorgon and Pluto LNG projects and $70 billion in coal-seam gas production. The major oil companies have proceeded with these projects without any fuss about excessive taxation.

The RRT is an Australian invention that is now regarded as best practice by institutions such as the IMF, the OECD and the World Bank. But even though our own economists invented it, the federal government was reluctant to extend it to onshore mineral production, as this is the states' domain. The work of Garnaut and Clunies-Ross did lead the newly formed Northern Territory government in 1982 to introduce a profits-based mineral royalty, which simply takes one in every five dollars of profit earned by its miners, but none of the states has since followed this eminently sensible model. Instead they've stuck with their anachronistic regimes, letting the established miners off the hook while discouraging small to medium-sized compaies. Royalties are a badly designed impost because they force both mature and emerging miners to hand over their pound of flesh *before* the operation has turned a profit. Typically, these royalties range from 6 to 10 per cent, depending on the type of mine. Developing countries tend to favour royalties because they like to get the money upfront, rather than waiting a few years to hit the big time. This is of course understandable for these countries, as they are often in dire need of money. They would prefer to have $1 million of revenue today, rather than wait a few years for $2 million. But it doesn't make sense for Australian states, which preside over mature economies with broad tax bases; they can afford to wait to cash in on mining profits. For thirty years our state treasurers have known—or should have known—of a more effective and efficient way of taxing the mineral sector, but as a result of sheer laziness or incompetence—or both— they have not acted. Yet again, Australia is operating like a developing country.

FOLLOWING THE MONEY TRAIL

Trying to establish just how much money the industry makes is no longer as straightforward as it once was. By and large, most people seem to think that it's a good thing if miners make a lot of money, whereas the same rule doesn't apply to banks, who cop a hiding every time they announce big profits. Perhaps this is because resource companies do a very good job of keeping the locals happy.

Mining spreads money around in myriad ways. As mining companies have relatively few workers compared to other industries, they can afford to pay their employees handsomely. In theory this circulates money through nearby towns, bolstering the local economy. In hard-pressed rural Queensland, which is emerging from a decade-long drought, the development of mines in the interior probably stopped some regional centres from imploding as farm incomes collapsed. The industry was the saving grace of several regional towns in that state. But increasingly, many miners are 'fly in, fly out' workers, commuting from major cities and towns. Some workers live in small rural towns in the eastern states and fly across the country to work fourteen-day shifts in the Pilbara. This means the money earned by a miner working in Karratha may be spent on the other side of the country.

Rather than build towns or rely on local labour (which is sometimes limited), many companies now fly, bus or drive their workforces in and out of operations, especially in remote regions. Rio Tinto and BHP built mining towns in the Pilbara in the '60s and '70s, but this ended in the 1980s. Now, about 5,000 of Rio Tinto's Pilbara workers fly there from outside the region. This means that local communities can experience the ill-effects of mining while getting very few direct benefits. In towns like Narrabri in northern New South Wales and the Queensland coal towns of Blackwater, Dysart and Moranbah, enormous work camps have been built on the outskirts of towns to accommodate the revolving labour force. Tony Maher, national president of the mineworkers union, CFMEU, says the FIFO model means mining companies are now contributing very little to the development of regional Australia.[84]

Cedric Marshall, mayor of the Isaacs regional council that covers key Queensland coal towns, says that in places like Coppabella the 1,700-bed mine camp is many times the size of the town, and in some bigger towns the camps are as big as the towns themselves. Marshall says the transient workforce means the towns miss out on economic benefits, while the extra traffic makes regional highways more dangerous. He says there are hundreds of trucks on the road each day, many of them with wide loads. The region used to transport its grain, general freight and fuel by rail, but this is all done by road now because of the demands of the coal industry on the rail network. A coal train now leaves the centre of Coppabella every twenty minutes.[85] A detailed study by Professor John Rolfe of the Central Queensland University found that 46 per cent of all the income generated by mining in the state, and 55 per cent of all contracts and community spending, flowed back to Brisbane via FIFO workers and contractors.[86]

In local communities, however, mining companies often take over where government is found wanting. Their big profits and relatively low tax rates allow them to operate as generous benefactors. They donate to local clubs and community organisations, and are often there in times of need. After floods, bushfires and cyclones, the big mining companies like BHP and Rio have made substantial dona-tions to relief funds. Country Fire Association units rely on donations, and thanks to mining companies they have some of the best high-tech gadgetry money can buy. Coal companies in the Hunter Valley support mobile libraries and home reading programs; primary school children read books emblazoned with the logo of a multinational coal company. In Orange, New South Wales, home of the Cadia Valley gold mine, local councillor Fiona Rossiter says the company has been a 'lifesaver for a number of projects'. She names two programs for disadvantaged children that had their funding cut by the state government; Cadia stepped in with a $500,000-a-year community program. 'They are quite giving. I realise that's because they make a lot of money and they have got plenty of profit, but they don't have to,' says Rossiter, a nurse and mother of eight children.[87]

The Cadia Valley mine, operated by Melbourne-based company Newcrest, is on its way to becoming the largest underground mining operation in Australia and the second largest gold mine in the world, under a $2 billion expansion plan approved by the state government. The local council and the state government were very keen to see the development go ahead, notwithstanding the city's perilously low level of water. The expansion plan was approved while Orange was on the maximum 'level 5' water restrictions, even though the mine draws on local water courses and ground water. But Orange now looks to be a reasonably prosperous place, with new shopping malls and rising house prices. There will be a lot more largesse in Orange over the next thirty years, the expected lifespan of the expanded mine.

As Mrs Rossiter observed, these companies are not short of money, but we know less about just how much they are making and where it goes. Both industry and government have cut back the amount of information reported to the public in recent years. Until the 2006–07 financial year, the industry produced the annual *Minerals Industry Survey*, which showed exactly how much it was earning. But the report was discontinued, perhaps because it was starting to show an embarrassment of riches. The Bureau of Statistics also scrapped its detailed biennial publication on the industry, *Mining Operations Australia*, which means information on payments to states by mining companies is no longer obtainable in a single publication.

After painstakingly splicing together information from multiple sources, two RBA economists, Ellis Connolly and David Orsmond, have made up for these deficiencies. They looked at information available from the Australian Taxation Office, from federal and state government budgets and from the ABS to show us where the money goes.[88] At the height of last decade's boom, wages cost $11 billion, or 9 per cent of total mining revenue, which is very low when compared to other industries, and obviously reflects the highly mechanised nature of mineral production. The latest figures show that the industry employs 194,000 people, up 90,000 since the start of the boom, but this represents a tiny 1.7 per cent of total employment in Australia. Given that the industry produces at least 10 per cent of our GDP, mining is six times more capital intensive than the average business.

The study also showed that spending on inputs like machinery, transportation and catering soaks up $45 billion, or 38 per cent of revenue. Most of this spending probably goes offshore, given the heavy use of imported machinery. The study doesn't quantify how much of these inputs are produced locally, but there are a few compelling indications.

In its corporate video for the Pluto LNG project, Woodside boasts that its steel fabrication work was done in China. Chevron's $43 billion Gorgon LNG project has been accused of sourcing just $3 billion of its construction contracts locally, contrary to company claims of $10 billion.

Ian Cairns, national manager of industry development for the Australian Steel Institute, an industry peak body, says Australian manufacturing has largely missed out on opportunities in the last five years. He says that all but 7,500 of the 260,000 tonnes of steel used in Chevron's Gorgon project were fabricated overseas. The figures for Chinese-backed projects are even more extreme. Hong Kong-listed Citic Pacific's Sino Iron project has had all of its 100,000 tonnes of steel fabricated in China, and it even imported concrete blocks from there, despite the prohibitive cost of shipping something so heavy. Under the federal government's Enterprise Migration Agreements the company will be able to import most of its workers on low wages, despite union warnings about the risks of using poorly qualified workers who may not even speak fluent English. Cairns says the federal government could use Foreign Investment Review Board (FIRB) conditions to boost local involvement, but he expects that the industry will get very little out of the $70 billion being invested in coal-seam gas projects. Martin Ferguson, however, is unsympathetic: 'It is not a gravy train to jack up your tender price thinking these investors have to accept your tender when you are eight to nine times higher than alternative bids from overseas. This argument that we can impose these sorts of additional costs on industry is rubbish.'

Before they make a profit, miners also have to pay royalties to state governments, which are the equivalent to a share of production. In this case, royalties are worth $7 billion, or 5.8 per cent of revenue. This leaves a gross profit of $56 billion, or 47 per cent of revenue, which is a very decent margin indeed. From this amount, the companies paid $8.1 billion in federal taxes, leaving a net profit after taxes and royalties of $48 billion. In this study, taxes and royalties as a share of net earnings amount to 26 per cent, which is definitely on the low side, especially given the boom conditions at the time.[89] This level of profit translates into a very high return on capital, ranging from 24 to 32 per cent in the three peak years of the mining boom last decade.[90]

The mining industry argues that its tax rate is higher than that shown by the RBA study. In recent years big miners like Rio and BHP claim they have been paying tax rates on their gross profits in the range of 35 to 43 per cent. A study by the accounting firm Deloitte based on Tax Office figures shows that in 2007–09 mining companies paid a corporate tax rate of 27.8 per cent, which rose to 41 per cent when royalties were included. In the decade to 2009, the industry says, it paid $80 billion in tax, or about 38 per cent of its pre-tax cash revenue of $210 billion.

Concern about the effectiveness of taxation becomes greater when very high foreign ownership is factored into the equation. Foreign ownership of Australian mining and energy businesses, now around 80 per cent and rising, means that some of these high profits leave the country, a trend that is likely to accelerate as miners deplete our natural resources and run out of new investment opportunities. BHP Billiton claims to be only 41 per cent foreign owned, but that figure is derived by counting its shareholders by number, not how much each of them owns. More detailed analysis by Thomson Reuters puts the figure at 76 per cent, a figure quoted by its rival Rio Tinto in a submission to a Senate inquiry. Rio put its own level of foreign ownership at 83 per cent, while Xstrata is 100 per cent foreign owned.[91] Official data on foreign ownership in the industry is now unavailable, after the ABS cut its annual publication on the topic in 1985. At the time (when foreign ownership of mining businesses was 50 per cent), it became unfashionable to produce such data.[92] Since then, major Australian companies like Western Mining and MIM Holdings have disappeared. Further analysis by Dr David Richardson of the Australian Institute, and Naomi Edwards, an economics adviser to the Australian Greens, puts overall foreign ownership of mining in Australia at more than 80 per cent. Edwards has pointed out in a briefing paper for the Greens that foreign ownership will rise given the soaring value of mining projects approved recently by the FIRB. The value of approved projects has quadrupled in the three years to 2009–10, reaching $75 billion.[93]

Both Edwards and Richardson argue that mining is three times more profitable than the non-mining economy, and they see these profits being shipped offshore in the form of dividends. Currently the foreign miners are reinvesting because digging our dirt is highly profitable, but it is conceivable that they will eventually recover their retained earnings and assets when Australia runs out of resources. Based on Tax Office returns for 2008–09, Edwards puts after-tax profits at 26 per cent of revenue for miners, compared with 8 per cent for other industries. Richardson used ABS figures for 2009–10 to produce a similar picture of after-tax profits: 37 per cent of revenue for miners versus 13 per cent for non-miners. Edwards's analysis shows that iron ore has become fantastically profitable, with after-tax profits at 48 per cent of revenue. These profit levels are generating handsome returns to foreign owners. Based on ABARES's forecasts, Edwards predicts that foreigners will accrue earnings of $265 billion over the next five years, more than half of which will be earned by iron ore alone. More than $50 billion will leave the country as dividends.

Edwards, a former actuary who sidelines as a stand-up comedian, can find nothing amusing about the implications of foreign control of our resource wealth. She argues that this is driving a deficit in Australia's net income balance, which is the difference between the earnings Australians make on their overseas assets and the earnings on foreign investment in Australia. Australia ran a positive balance until the mid-1980s, when it dropped to a deficit of about 2 per cent of GDP for the next two decades. The deficit is now heading for 6 per cent, or about $75 billion in net payments to overseas investors. Edwards isn't raising problems about foreign ownership per se; the real issue is the nature of the mining business. If these companies controlled resources that were renewable, foreign ownership would be less of a concern; but the fast and highly profitable rate of resource extraction, combined with the lack of local asset development, threatens to leave us a greatly depleted nation when the resources eventually run out. Like Nauru today, but on a continental scale.

The boom that is now underway is potentially going to be bigger, longer and more profound than any of the previous four booms in Australia's history. Without a more effective tax regime, Australia won't be able to reap a legacy for the lasting benefit of present and future generations.

Raiding the Food Bowl

'*You'll never understand the love of the land. Think of it like this—I love the land as much you love money.*'—Tim Duddy, sixth-generation farmer, in a meeting with Scott Sullivan, BHP's New South Wales director of coal, in 2010[94]

After more than a century of industrial mining in Australia, there are some deep and permanent scars left on this continent. Strangely enough, a lot of people think they are pretty impressive. Kalgoorlie's Super Pit is so wide and deep that it can be seen from outer space and has become a tourist attraction. Broken Hill now has a slag heap the size of a small mountain; tourists can drive to the top, eat at an upmarket restaurant and take in a spectacular sunset. And in western Tasmania there is an entire valley around the old Mt Lyell mine that looks like a moonscape. The locals are proud of it.

Most people probably think such scars are a small price to pay for the wealth that has been extracted by mining, especially in these remote areas. But this sort of damage is only the beginning. New technology, soaring mineral prices and massive investment are combining to produce projects on a scale that is barely imaginable, raising the threat of serious environmental consequences for future generations. We get the benefit; future generations cop the enduring cost.

Some proposals recently approved or now before government show how resource companies are pushing the boundaries of technology and regulation like never before. BHP Billiton plans to increase the size of its Olympic Dam mine in South Australia by 500 per cent. It will leave behind a toxic lake and a 44-square kilometre mound of radioactive tailings, which will remain active for 10,000 years. The Queensland government has confirmed that coal-seam gas projects are likely to involve the drilling of up to 40,000 wells on farmland in the state's southeast. Advice to the federal environment minister obtained under FOI reveals that the projects involve a host of ecological risks and uncertainties about water consumption, land subsidence, waste saline water and the process of 'fracking', which involves pumping toxic chemicals, sand and water into the ground under extreme pressure to release gas. In Western Australia, Woodside wants to build an LNG plant amid the pristine marine environment at James Price Point, even though other locations present far less risk to the Aboriginal and environmental heritage. And the British multinational BP, now infamous for creating the worst oil spill in US history, has been given approval to drill in deeper and more treacherous waters in the Great Australian Bight. BP's permit area plumbs depths of 4,500 metres, three times the depth of the spot where the Deepwater Horizon rig blew up in April

2010 before spewing 5 million barrels of oil into the Gulf of Mexico over eighty-five agonising days.

No longer is it possible to consider the impact of a single mine in isolation. From the 1970s onwards, all mining and energy developments have been subject to environmental assessment, typically known as an Environmental Impact Statement. But this process is becoming redundant now that mines tend to be built in clusters, leading to what experts call cumulative effects. In the oldest coal-mining region in Australia, the Hunter Valley in New South Wales, dust and heavy metals are found in the air in high concentrations and serious health consequences have emerged. The Hunter Valley is home to more than thirty open-cut coal mines and is likely to get a whole lot more, given the proposals now before government. Satellite images show that between the towns of Muswellbrook and Singleton, a triangle measuring 30 by 50 by 50 kilometres contains as many as twenty exposed pits. One pit, just 6 kilometres west of Singleton, is 12 kilometres long and 4 kilometres wide in parts. Immediately to the west is farmland, including a macadamia farm. In central Queensland a string of more than thirty coal mines runs along the western edge of the Great Dividing Range east of Mackay, Rockhampton and Gladstone.

Chloe Munro, the national water commissioner, says the scale of mining developments has serious implications for water resources and entails unknown risks for water consumption and quality. 'One development on its own is one thing; when you have a whole series of them the cumulative effects can be quite dramatic,' she says. Munro is calling for coal-seam gas projects to be included in the water allocations under the federal–state National Water Initiative (NWI). Currently they are covered by 'special clause 34', which excised mining from the initiative when it was first signed by Prime Minister John Howard in 2004. The scale of new developments and the uncertainty surrounding their projected water consumption is alarming, Munro says:

> We are concerned those provisions are being stretched too far just because of the scale of the impact on the water resource that coal-seam gas extraction can represent. That is why we are calling for a precautionary approach. It is of concern to us both in terms of volume of water and the potential quality impacts. The management regime for that isn't as robust as it might be.[95]

Munro says mining was excised from the NWI because many projects were of short duration and used low-quality water which had no alternative use. Rapid expansion, however, means that mining is now taking place in areas where water planning has not been done. The National Water Commission's 2010 Mining Position Statement says clause 34 is not being applied in a 'consistent and transparent way'. The paper essentially argues that the clause should no longer apply. 'Wherever possible, mining activities should operate under the same rules and regulations as everyone else.'[96]

Mining uses water to process ore and to control dust. In the Hunter Valley, the state government has responded to community concerns and installed fourteen air-quality monitoring stations following reports in the local paper, the *Singleton Argus*, about high levels of respiratory problems among children. But there are concerns that these monitors might not be measuring the particles with the most serious health consequences. Only three of the monitors will measure ultra-fine particles known as PM2.5 (or 2.5 microns), while the other eleven will measure larger particles known as PM10 (10 microns). Dr Wayne Smith, the director of environmental health for the New South Wales Department of Health, who is also a professor of epidemiology and the son of a coal miner, admits that the smaller particles might be the real concern. He says more PM2.5 monitors were not installed because Australia is yet to agree on a standard for safe levels of these particles—another area where the growing size of the mining industry makes existing regulation deficient.

> There are developing guidelines for PM2.5. There is some evidence from overseas that the smaller particulates potentially have worse health effects. That is not conclusive evidence, but there is enough there for people to say, 'Yeah, this is probably an issue and you probably should be measuring smaller particulates.'[97]

Dr Smith says that after looking at exhaustive data on the health of people in the Hunter Valley, he thinks asthma rates might be higher, but the results are not statistically significant. Dr Dick van Steenis, a retired British physician who has studied the health effects of mines and power stations in the UK, shares Dr Smith's concerns. Dr van Steenis has been invited several times to come to Australia and speak to communities about the adverse effects of mining. His talks in these mining towns have typically drawn big crowds of interested locals. He argues that people living within 5 kilometres of a mine have an increased risk of suffering asthma, heart disease, stroke, type-2 diabetes and of having underweight babies. He says Australia's regulations and monitoring are well behind overseas standards. The monitors in the Hunter Valley should all be measuring PM2.5s and related polycyclic aromatic hydrocarbons (PAHs), which are generated by the heavy diesel-powered machinery used in mines and by transporting coal. The release of these particles could be minimised by mandating particle filters and by using foam to cover coal when it is transported and unloaded at the export terminal. While the states have begun looking at this issue, the federal health department appears to have ignored it. An FOI request for advice on the dangers posed by PM2.5s, and the need to develop national standards, turned up no documents during the two years to March 2011.

BLACK SOIL, BLACK GOLD

The cumulative effects of mining at Broken Hill contributed to a searing, life-changing moment for Peter Andrews when he was just three and a half years old. The year was 1943 and Andrews's family was running a sheep farm in the area when an almighty dust storm ripped through the region. The family survived by bunkering in a cellar, but many of their

3,000 sheep perished. Throughout his life, Andrews has thought about the practices that led to such devastation. Agriculture was a contributing factor, but so was the Broken Hill mine, which denuded the surrounding region of vegetation. For hundreds of kilometres around the town, thick stands of cypress pines, leopard woods and casuarinas were cut down to feed the development of the mine and the city of Broken Hill.

Andrews went on to become a guru of sustainable agriculture. He moved to the Upper Hunter region and turned the over-grazed country on his Bylong Valley property into lush, productive farmland. He has done this without the benefit of irrigation or chemicals—and by letting weeds grow. He has twice been featured on the ABC's *Australian Story* and is asked to share his innovative techniques all around the country. This should be a story of agrarian nirvana, but mining has again caught up with Andrews. For most of his life in Bylong, he and his family have lived in the shadow of mining. The Bylong Valley sits between the coalfields of Singleton and Mudgee. A railway line runs through the valley to the Hunter Valley export terminal. And now the Andrews family has learned that there is coal underneath their black soil and a Korean company has a licence to carry out exploration work.

When weak, inept and corrupt governments get into bed with powerful mining interests, the results are disastrous for the people at the coalface. In New South Wales, the disgraced former Labor government plastered the state with mineral exploration leases during the sixteen years it was in power.

Some of these leases were issued without any tender process. The number of coal leases doubled during this period and mining and petroleum leases now cover about half the state, compared with 21 per cent in Queensland. The leases cover land that people never ever thought would be mined, like the Andrewses' Tarwyn Park property, the black-soil Liverpool Plains, rural towns throughout the state, vacant industrial land in inner-city Sydney, the seabed off the coast of Sydney and even the Southern Highlands rural retreat owned by Nicole Kidman and Keith Urban.

Thousands of farmers are beside themselves because they see the real threat of having to sell their properties to the miners once they become enveloped by pits. The mining boom has turned farmers like Andrews's son Stuart into political activists. Stuart, who looks so typically rural that he could appear in an R.M. Williams catalogue, has joined the local Bylong Valley Protection Alliance. Stuart Andrews says his land could be productive for thousands of years. He cannot understand why Australian governments would allow such farmland to be sacrificed for short-term gain. The same goes for Tim Duddy, whose 5,000-acre spread at Caroona, on the ridges of the Liverpool Plains, is overlapped by a BHP Billiton lease. For seven years now, the irrepressible Duddy has been a full-time farmer and a full-time political activist, exploiting all the connections gained from his years as a boarder at an elite Sydney school to win the fight. In 2008 he staged a 615-day blockade against BHP's access to drill exploration bores on his land—a right the company has under law. Duddy's blockade led to a Supreme Court ruling in early 2010 by Justice Monika Schmidt, which said the previous land-access agreement was 'invalid' because BHP had not properly consulted all landholders (including the banks) and had not addressed environmental concerns. Other farmers are starting to adopt Duddy's 'lock the gate' approach to mining encroachment, especially those faced with coal-seam gas developments.

Across the valley from Duddy is a lease controlled by the Chinese mining giant Shenhua. In 2011 the journalist Natasha Bita revealed that since forking out $300 million to the New South Wales state government for the lease, Shenhua has spent more than $200 million buying up forty-three farms around Gunnedah in the hope that the state government will approve its development.[98] Given the serious money it has outlayed, Shenhua is obviously counting on getting the green light. It has already bypassed Australia's Foreign Investment Review Board because the deals fall under the $231 million threshold. Duddy stresses that the locals are not opposed to the development because it is Chinese, although he says some are alarmed to see a Chinese flag flying outside Shenhua's head office in Gunnedah. The federal government was forced to admit that it doesn't know how much land has been acquired by foreign interests for mining and has now asked for an audit.

Leaving aside legitimate concerns about allowing the Chinese state to own Australian land, what really concerns the people of Caroona is the damage they are already seeing, even before mining is approved. Underneath the Caroona Valley and the Liverpool Plains are about 50 metres of rich black topsoil, and underneath that is a series of aquifers. Locals fear these will be irreparably damaged if mining goes ahead. Even drilling exploration wells poses some risks to these structures. Duddy argues that while BHP has lifted its game and is now adopting best-practice with its exploration work, Shenhua's drilling has been invasive and the company has failed to notify farmers of its plans. 'Now we wake up to see a drilling site erected within metres of our boundary fences with no consultation, notice or discussion. Shenhua has totally failed to understand that in an area where we are subject to water-sharing plans, you may well destroy the livelihood of the people over the fence if you do not follow proper procedure,' he says.

The experience of one Upper Hunter farmer, Fiona Nevell, foreshadows what might be in store for the people of the Liverpool Plains.[99] In the forty-seven years that Nevell's family have lived on their Wilpingjong property, they have always been able to drink pure spring water from a source up the hill—until early 2010. Nevell knows the history well: her family bought the farm from descendants of the original settlers. The spring was first discovered in 1853 and the water was considered so good that a farm manager used to send workers to collect it in buckets for his whisky. The Nevells began piping the water to their home soon after they bought the property and were still drinking it until recently, when it turned brown. 'I think those people mucked up my water. It has all become sludge,' says Nevell. Asked if she has considered asking for compensation from the nearby mine, she says: 'I'm a farmer—we don't get compo.' The owner of the mine is the US giant Peabody, which mines 240 million tonnes of coal and earns about US$7 billion in revenue a year.

RISKY FRACKING BUSINESS

The Liverpool Plains farmers may soon have more to worry about than coal mines. Petroleum exploration leases, which could facilitate the production of methane gas from coal seams, cover 29 per cent of New South Wales, stretching from the Illawarra region to Sydney, through the Hunter Valley and all the way north to the Queensland border. These leases join CSG developments in Queensland that have spurred $70 billion in investment for three projects alone. The Queensland CSG projects will involve a vast net-work of wells and pipelines stretching across more than 17,000 square kilometres of prime farmland, from the Bowen Basin west of Gladstone, all the way south past Maryborough to the Surat Basin and on to the Darling Downs west of Brisbane. One project alone involves 10,000 kilometres of pipelines and access roads to harvest gas from 10,000 wells. The inland rural centres of Chinchilla, Dalby, Roma, Tara, Toowoomba and Wandoan are in the thick of frenzied development as these projects race to supply export contracts. The Queensland Department of Employment, Economic Development and Innovation says that should development plans hit the 'mid-range' of about 28 million tonnes of gas per annum, 20,000 wells will be drilled over the life cycle of these projects. If all eight known projects reach full capacity, this figure would rise to 40,000 wells.

So far the state and federal governments have approved three projects, with two already under construction: QGC, a subsidiary of the UK multinational BG Group, is building the $16 billion QCLNG project involving 6,000 wells that will pump gas to an LNG processing plant; and the $15 billion Santos-led consortium, Gladstone LNG, will drill 2,650 wells. Both projects will build LNG plants and export terminals on Curtis Island, one of the five major coastal islands within the Great Barrier Reef World Heritage Area. These developments have put a rocket under house prices in the towns, but farmers who already have wells on their properties say their land values have fallen. One landowner who has been waging a war against QGC is disability pensioner Paul Keating, whose 50-hectare property at Tara, west of Toowoomba, already has wells and pipelines on it. Mr Keating has tried to block access to his property after claiming that QGC was building an access road wider than agreed in the contract. Now he has had enough and wants to sell, even though this would involve taking a deep discount. 'People have come out to inspect the property and they love my house, but as soon as they find out it's affected by the QGC pipeline they're gone,' Keating says.[100]

These two projects are so big that they rival the size of Australia's current LNG industry, but they are just the beginning: the third proposed project is more than double the size of these two combined. Australia Pacific LNG (APLNG), a $35 billion joint venture between Origin Energy and US giant Chevron, has federal and state approval to build another LNG plant and to drill 10,000 production wells covering an area of almost 6,000 square kilometres of farmland. A fourth project, yet to be approved, is Arrow CSG, a joint venture between Shell and PetroChina. At 16 million tonnes of LNG a year this project would be on the scale of APLNG and would likely require 10,000 wells. Together, these projects involve investment of almost $100 billion and will treble current LNG production.

Advice to the federal government on these projects reveals that decision-making has paid no regard to cumulative enironmental effects. The rationale for federal approval of the APLNG project was that its risks were similar to those of the first two and that it should therefore be approved. The environment minister, Tony Burke, was advised by the Department of Sustainability and Environment (DSE) in February 2011:

> On the whole, these uncertainties and likely impacts are largely similar to those for the Santos and QGC coal-seam gas projects, which you approved under the EPBC [Environment Protection and Biodiversity Conservation] Act on 22 October 2010. As such, we have recommended relatively similar conditions to manage the uncertainties and likely impacts as the conditions of your approval for those projects.[101]

But the risks, uncertainties and unknowns outlined by the department concerning the project are considerable, and these are likely to grow as more coal-seam gas projects come on stream.

On groundwater, the submission to Burke cites advice from GA that warned of 'high levels of uncertainty in the predicted impacts of CSG development on groundwater behaviour and on EPBC listed ecological communities'.[102] Burke approved the project even though he was told that GA believed 'APLNG's modeling requires further work to fully establish uncertainties'.

On the use of chemicals to 'frack' the coal seams and release gas, the department revealed that the federal and Queensland governments withdrew a requirement to have zero levels of highly toxic BTEX chemicals (benzene, toluene, ethyl benzene and xylenes) after the company protested. Turning coal-seam gas into LNG for export involves blasting a combination of toxic chemicals, sand and large amounts of water into the coal seams below the surface to release gas. This process has serious consequences for the watertable. The National Water Commission is not just worried about the amount of water used by CSG projects; it also has real concerns about the impact on water quality. Chloe Munro cites the case of Cougar Energy, which had to stop work after it was discovered that it was using BTEX. Traces of the same chemicals have been found in the APLNG project. In the department's February submission to Burke, it reported that the Queensland government had met with APLNG. The department subsequently advised Burke's departmental secretary that there were 'no impacts on landowner bores and no evidence of environmental harm from this incident'. This briefing note indicated that the Queensland government accepted at face value what APLNG told it about the levels of BTEX. 'From information provided to the department by APLNG, the findings of BTEX chemicals are very low … in most cases below levels for BTEX chemicals in the Australian Drinking Water Guidelines.' Both governments then agreed to remove a zero BTEX requirement from their list of conditions because 'it may present practical difficulties to implement in advance of broader regulation to address this issue'.

On salinity, Burke was warned that these projects would bring to the surface massive amounts of water containing 'high salt concentrations and … a high sodium absorption ratio, which means it would be likely to cause environmental harm if it were released in significant volumes to land or surface water'. Over thirty years, the APLNG project alone is expected to bring '2 million tonnes of chemical salts (comprising a range of chemical components/contaminants)'.[103] The paper also notes that 'APLNG does not have definite proposals for the management of CSG water on the surface'. This water could damage farmland in the event of a massive flood and could also threaten the Narran Lakes wetlands, located about 250 kilometres from the project. The department said there was a risk of 'large-scale flood events, as seen recently in southeast Queensland, which could overtop the brine storage basins thereby mobilizing salts and associated heavy metals downstream and into the Narran Lakes'. The department noted that this is an 'internationally significant area for waterbird breeding'.[104]

On the wider impact on the farmland of the region, Burke was warned of the risks of building 10,000 kilometres of pipelines and access roads and tracks. The department said the location of this infrastructure 'has not been determined with certainty, and the size of the gas fields area means that it is not possible to identify detailed environmental impacts with precision'. An earlier submission on the first two projects said 'the proposals are also likely to cause land subsidence'.

One risk not addressed by the department in its advice was climate change, which is forecast by international and local experts to make southern Australia considerably drier, with much longer droughts.[105] Given this scenario, it seems untenable to allow developments that will massively increase water consumption. The CSG industry will be using 'a very significant amount' of what Chloe Munro calls 'dinosaur water' from the Great Artesian Basin. A National Water Commission position paper says that these projects are likely to use about 300 gigalitres (GL) per year, an increase of 55 per cent above the 540 GL currently drawn from the Basin, and this is their conservative estimate.[106] But estimates contained in a July 2010 letter written by the then environment minister, Peter Garrett, and obtained under FOI said that 18,000 production wells could extract '15,000 to 28,000 GL (possibly up to 45,000 GL) of groundwater'. The note added: 'The proposals will also affect a significant number of stock and domestic users who rely on groundwater in the region, and have the potential to cause widespread subsidence on the land surface'.[107] If these estimates are realised, extraction rates would be 1,000 to 1,500 GL a year, or two to three times higher than the current rate.

The government's assessment of the water impacts of this project followed a new requirement, introduced by the independent MP Tony Windsor, that all such projects be subjected to an 'independent expert study'. The study was completed by Professor Chris Moran, who heads the University of Queensland's Sustainable Minerals Institute (SMI). Moran also completed the water study for the QGC project and the groundwater risk assessment for the Queensland government. SMI's estimate, based on publicly available information, is a low 200 GL a year, but Moran says a lot more research needs to be done. Estimates of the groundwater impact of these projects vary considerably, he notes, because each aquifer is different. Conservationists have questioned SMI's involvement in these studies, and Moran confirms that it receives 60 to 70 per cent of its funding from the industry. Its mission statement says it engages in 'industrially applicable research'. SMI used to produce an annual report that disclosed its industry links but no more, perhaps as a result of a 2007 paper by Dr Clive Hamilton citing it as a prime example of 'university capture' by industry.[108] Even now, however, all of the 'related links' on SMI's website are mining industry associations or research centres. Moran defends his record, declaring that 'industry cannot purchase results'.[109]

The ability of companies and governments to monitor and regulate projects of this scale seems questionable. When Tony Burke granted federal approval in October 2010 for the first two CSG projects he said he had imposed strict conditions concerning maintenance of water pressure in aquifers. But Queensland's former principal hydrologist, John Hillier, says that leakages may not be detected for two, three or even four decades, perhaps long after the gas has been extracted and the project decommissioned. Hillier believes mistakes will be made in the rush to commission up to 40,000 wells and suspects that as many as 5 per cent, or 2,000, will prove faulty. In his day, inspectors were required to attend the drilling of every bore in the Great Artesian Basin in Queensland, but the gas rush has made this impossible. Some farmers have already reported explosive gas leaks from wells on their land. In May 2011, one of Arrow Energy's wells blew up on farmland near Dalby, 200 kilometres west of Brisbane, sending a shaft of gas and salty water 90 metres into the air.[110]

Farmers affected by these developments believe the state government is blinded by the royalties it expects to receive and will not impartially assess the threats to water resources. Cotton farmer Graham Clapham, from Queensland's Toowoomba region, accuses the government of regulatory bias. 'There should at least be an independent third party to regulate the environmental conditions under which the companies operate. It should not be the same body that is going to receive the royalties.' Clapham is one of a group of farmers challenging the Queensland government in the Land Court over its ability to protect environmental values in dealing with Arrow Energy's project. Another is Ruth Armstrong, who is driving the court challenge. As she told the *Australian*'s environment editor, Graham Lloyd, 'We are challenging the conditions of the environmental authority with regard to the impact on water, soil and lifestyle. We think the existing authority is inadequate to protect those environmental values.'[111]

Queensland's CSG developments raise questions about the capacity of state governments to regulate such projects effectively. It is no wonder that more farmers in New South Wales and Queensland are opting to blockade exploration on their land. Conservationist Drew Hutton launched the 'Lock the Gate Alliance'; it now has eighty regional groups and will possibly mount a High Court challenge to the rights of miners to enter freehold farmland.[112]

The limited capacity of smaller jurisdictions to manage these developments was brought into even sharper focus with the 2009 Montara oil spill in the Timor Sea. The ten-week spill exposed breathtaking shortcomings in the performance of the Northern Territory regulator, which accepted at face value what the Thai company PTTEP was telling it. The explosion of the Montara rig in the relatively shallow waters of the Timor Sea led to Australia's worst ever oil spill. Since then, Martin Ferguson has bolstered national regulation so that the National Offshore Petroleum Safety Authority (NOPSA) now has responsibility for the integrity of wells. NOPSA has become a de facto national regulator for the offshore industry. But this is just a stop-gap measure: Ferguson wants to go further and introduce a national regulator, to be called the National Offshore Petroleum Safety and Environmental Management Authority, to regulate the entire offshore industry. The Western Australian government, however, is refusing to join in.

Ferguson says the takeover of day-to-day regulation of the offshore industry began nearly thirty years ago, when the High Court ruled that the Commonwealth had exclusive jurisdiction in Australia's offshore waters. It is unlikely that the federal government could ever get control of onshore regulation. 'There is no head of Commonwealth power relating to the regulation of natural resources and accordingly, constitutionally, the states are responsible for regulating onshore coal-seam gas operations,' he explains. But Chloe Munro points to a third way. She says that while a federal takeover would mean 'you are more distant from the on-ground information', joint regulation would have many advantages. 'There could be a pooling of resources and people could stand shoulder to shoulder as they make changes. That is the strength of the National Water Initiative, because it is agreed by all parties. A collective rather than a centralising thing,' she says.

In fact the NWI is a good example of how state and federal governments can pool resources to beef up regulation. Clearly there's a case for this happening with coal-seam gas and a raft of projects that are about to bounce off the drawing board. More recently, the federal government showed how it might play a role when it partly funded the Namoi Water Study, following concerns about mining in the Liverpool Plains. But the initiative comes with a sobering twist. The Feds contributed $1.5 million to the study, but most of the money actually came from the coal companies, who were able to stump up twice that amount. The companies would only have made a financial contribution if they thought they could influence the outcome.

MAD MAX MINING

BHP Billiton's Olympic Dam mine in the arid country 560 kilometres north of Adelaide is already a very big under-ground mine—the largest such mine in Australia. It has the world's largest deposit of uranium, the world's fourth largest copper deposit, the world's fifth largest gold deposit and significant quantities of silver. The existing mine is a highly mechanised operation, but it is still so big that it requires 1,500 BHP employees and another 1,500 contractors, making it one of the single biggest mining operations in Australia. The mine, which has its own smelting operation, currently extracts about 10 million tonnes of ore a year. It commenced operations in 1988 but, as one manager says, this 'is just the start'.

BHP has put a plan before the state and federal govern-ments to expand this mine dramatically by turning it into a mammoth open-cut operation that will run parallel with the underground mining. The expansion, expected to cost $20 billion, would take eleven years to complete and produc-tion would then run for at least thirty years. The expansion plan involves increasing annual ore production by 500 per cent to 72 million tonnes. Copper concentrate production would increase three-fold to 2.4 million tonnes; refined copper production would increase almost three times to 750,000 tonnes; uranium oxide production would increase about four-fold to 19,500 tonnes; gold production would increase seven-fold to 750,000 ounces; and silver production would increase 2.5 times to 2.9 million ounces. As a result of BHP's win over the super-profits tax, not a single cent of any above-normal profits would be returned to Australian citizens. And only 10 per cent of the expanded operation's smelting would be done at Olympic Dam; the plan involves shipping uranium-infused copper concentrate and uranium oxide to China for processing, even though it is a nuclear weapons state. This decision means that BHP is also shipping jobs to China.

The infrastructure required to build this operation gives some indication of the energy intensity of mining operations. BHP concedes the project could increase South Australia's greenhouse gas emissions by up to 9 per cent. The expansion will require the following new infrastructure: a desalination plant to supply 200 megalitres of water a day to the mine; a 270-kilometre power transmission line; a 100-kilometre rail line; new port facilities in Darwin and at Outer Harbour north of Adelaide; a new workers' village; and a new airport to fly in the FIFO workers on Boeing 737-800 or Airbus A320 jets. The company's corporate video says it is looking at renewable energy sources to run the desalination plant, but it won't commit to a minimum renewable energy target for the entire operation.

Aside from the immediate environmental impact, the Draft Environmental Impact Statement (EIS) shows that the Big Australian is becoming increasingly brazen as it plans to leave behind a radioactive legacy that is unprecedented anywhere in the world. After forty years of mining, the open-cut pit would be 1 kilometre deep and more than 4 kilometres wide and a dump of radioactive materials 'capped' with rock would be left nearby, covering an area of 44 square kilometres—the equivalent of 2,200 football fields, or bigger than the Melbourne CBD. BHP's own EIS indicates that it is proposing third-world standards for waste storage in Australia. A table showing storage techniques used by uranium mines around the world lists Namibia as the only other country that has allowed above-ground storage for such waste.[113] This massive pile of tailings will reach a height of 65 metres—just 2 metres shorter than the Sydney Opera House. Leaving the radioactive tailings above ground is also contrary to established practice in Australia. Rio Tinto's Ranger uranium mine in the Northern Territory is required to return its tailings to the pit and cover them completely. Commonwealth regulations for Ranger require that all tailings must be disposed of so that they are physically isolated from the environment for at least 10,000 years.[114] BHP argues that it has taken a different approach because Ranger receives ten times the rainfall received around Olympic Dam. The Olympic facility is guaranteed to remain stable for 1,000 years, the company claims. Let's hope so: the mine might be located in the arid country where the film *Mad Max* was made, but it is still only 560 kilometres from a capital city and about half that distance to regional centres like Port Augusta and Whyalla.

Meanwhile, all of the water produced by the operation's desal plant will have to go somewhere after it has been used to process the minerals, and the Draft EIS explains exactly where it would end up. Up to 8.2 million litres would seep from the radioactive tailings storage into groundwater *each day* in the first ten years of operations, before falling to about 3.2 million litres a day. The mine's impact on groundwater would 'depend' on various unknown factors, such as 'the inter-action of seepage with the sediments beneath the respective facilities', but the draft confirms that the seepage would cause a 'groundwater mound' below the tailings storage facility that would affect groundwater levels 'for up to 6 kilometres'.[115]

The exposed pit left by the mining would 'remain as a permanent land feature' and contain 'a wide range of concentrations of metals', including groundwater flow of 1 million litres per day from two local aquifers.[116] BHP's corporate video claims that its modelling shows the pit won't overflow, and that it has even taken climate change into account. It says groundwater inflow would remain 700 metres below ground level and not reach the top of the crater because of evaporation. After 1,000 years, a salt crust would cover the surface of the lake. But the operation would leach water into the watertable below.

BHP's plan would set a new precedent by exporting uranium in concentrate form, rather than as fully processed yellow cake, thereby potentially compromising Australia's uranium safeguards regime. Foreign Minister Kevin Rudd has confirmed that Australia is negotiating with China to amend its nuclear transfers agreement in order to facilitate the export of uranium-infused copper concentrate. The government has agreed to amend Annex D of the Australia/China nuclear transfers agreement.[117]

In a state like South Australia, which has lagged behind other states in the post-tariff world, proposals like this one have political leaders salivating, so much so that they are prepared to allow the company to operate under its own special act of parliament. To speed up development, the expansion would operate under the *Roxby Downs Indenture Ratification Act 1982*, which was passed to expedite development of the mine by Western Mining in the 1980s. BHP says it is still required to operate under 'all other relevant Commonwealth and State legislation', but the Australian Conservation Foundation (ACF) says this act confers special legal privileges and takes precedence over the state's modern environment protection regime. The minister responsible for the project, Kevin Foley, was unable to answer a series of questions for this book about the state's capacity to monitor and regulate the new development. 'The specific questions you have raised are being addressed as part of the ongoing consultations with BHP Billiton on the SEIS as well as the parallel negotiations on the amendments to the Indenture,' a spokesman for Foley said.

The ACF's chief executive, Don Henry, says he has seen a lot of bad proposals in his thirty years in the environmental movement, but he believes the Olympic Dam expansion is quite possibly the worst. The audacity of BHP in putting the proposal before state and federal governments speaks volumes about the sort of influence the company thinks it has, he says. Henry speculates that after its disastrous Ok Tedi mine in Papua New Guinea, which led to catastrophic environmental damage and prompted local landowners to bring a class action against the company, BHP would not dare put such a proposal in a developing country. But in its own backyard, the Big Australian thinks it can throw its hefty corporate weight around. Henry explains:

We are deeply worried about Olympic Dam, the extent of the degradation and toxic tailings they are proposing to leave behind, right through to being prepared to operate under a South Australian law that bypasses modern environmental legislation. I am sure the board does not fully realise they are putting their social licence to operate in jeopardy. It is certainly surprising that they seem to be running a high risk with their reputation in a way that's similar to Ok Tedi.[118]

BHP says that not leaving tailings above ground is 'best practice' for Olympic Dam because of its scale and the negative environmental and economic impacts of attempting to return excavated material to the ground. In other words, this mine will have to extract an enormous amount of earth before it reaches the ore, generating significant greenhouse gas emissions along the way. Putting this material back into a pit would double those emissions. It argues that those mines that do put tailings back underground are 'uranium only'. BHP also says that any comparison between Olympic Dam and Ok Tedi is 'ludicrous', given the difference in rainfall between the two sites. The company says it has spent $30 million on environmental studies to ensure that the environmental impact would be 'minimal'.[119] In what some conservationists believed was a Black Friday joke, BHP released its response to the 391 'unique submissions' made in response to the Draft EIS on Friday, 13 May 2011. This Supplementary EIS, the size of several telephone books, endorsed the key elements of the original proposal, including above-ground storage of radioactive waste and the export of uranium oxide to China. It reconfigured the tailings-storage facility, adding another 80 hectares to its expanse. The only major concession to public concerns involved building a tunnel for the outflow pipe from the desal plant, rather than digging a trench.[120]

Another interesting example of the trade-offs involved in these projects is the company's decision to abandon its original plan to transport ammonium nitrate via rail from Newcastle. 'Following discussions with the South Australian government, the transport of ammonium nitrate is to be by road from its point of origin to Olympic Dam,' the company explains. The EIS indicates that BHP might build the necessary rail spur at some point in the future, and shipments might then go by rail subject to South Australian government approval. This change means that 'security sensitive' material will initially be transported on the roads, adding up in the first five years to 1,900 annual double road trains, shipping 76,800 tonnes of explosive materials along the Barrier, Princes and Stuart highways. The company admitted euphemistically that there was an 'associated risk profile' resulting from this decision, but it seems both BHP and the state government think such risks are a small price to pay to secure a 'world class' resource development.[121]

Future generations who are left to live with the consequences of this mine may wonder why we needed a project like this. Those left to deal with the environmental costs of coal-seam gas expansion will wonder why we were in such a rush to take on so much risk and uncertainty, especially given our failure to compensate them for the damage. These projects not only have serious environmental consequences; they also come loaded with implications for governance in Australia for years to come.

Yours and Mine Country*

'Families have split up. People are not talking to each other, fighting in the street. It just breaks my heart.'
—Yindjibarndi elder Bigali Hanlon, seventy, on the effects of Fortescue Metals' divisive tactics in iron-ore negotiations, April 2011[122]

The people who have been confronted by mining more than any other group in Australia are Indigenous communities living in remote regions. The Minerals Council of Australia says 60 per cent of the mining in this country 'neighbours' Aboriginal land—or is often *on* Aboriginal land.[123] It would be nice to think that mining has improved the welfare of our poorest and most disadvantaged citizens, but so far such benefits have proved elusive.

Some of their experiences recall the more spectacular examples of the resource curse found around the world. It is as though pockets of Nauru or Zambia have been dotted around the remote outback. In these places, resource bonanzas have triggered a downward spiral accelerated by poorly

* Chapter title adapted from Scambary, B. *My Country, Mine Country: Aborigines, mining and development contestation in remote Australia*, PhD thesis, Australian National University, 2007.

targeted compensation. Resource wealth has sparked conflict between and within communities and families. Despite the lessons from places like Bougainville, where mining led to a decade-long civil war and a UN-estimated death toll of 15,000, some companies are still deliberately dividing communities in what look like old-fashioned land grabs. So far Australia has been spared the kind of violence experienced by the people of Bougainville, although that's not to say it could not happen. There is a large and growing number of aimless and angry young Aboriginal men living in the shadow of booming mining complexes.

The Northern Territory's Groote Eylandt is one such home-grown Nauru, while damaging effects of mining have also been felt in communities near the bauxite mines on the Gove and Cape York peninsulas and in Western Australia's Pilbara region. Mining companies are now trying to make amends, but they have a lot of catching up to do. Many of these communities have been living with mining for half a century. Three generations on, they typically suffer from very low rates of education and high rates of crime, drug abuse and incarceration.

Some major players, including Rio, BHP, Woodside and smaller companies like OZ Minerals, have made determined efforts to improve Aboriginal welfare. Many companies operate pre-employment programs to help Aborigines overcome their lack of education and work experience. Others, however, have shown little interest in the problem. The multinational Chevron has no specific program to employ or train local Aboriginal people living near its $43 billion Gorgon development, the single biggest resource project in the nation's history. The federal government doesn't seem to think this is of any concern.

When it comes to permission to access Aboriginal land, multinationals are still securing special authorisation when legal procedures get in the way. In the remote Gulf Region of the Northern Territory, Xstrata's McArthur River Mine (MRM) involved diverting an entire river a distance of 5 kilometres to allow for open-cut mining near Aboriginal sacred sites at McArthur River, ten hours southeast of Darwin. When the Federal Court ruled that the approval process for the river diversion was flawed, the Northern Territory government rushed through legislation to nullify the court's decision. Federal approval was then granted by the then environment minister, Peter Garrett. This didn't happen thirty years ago; these events took place in 2009. Even after these extraordinary interventions, the company found cause to complain. Xstrata Zinc's chief operating officer, Brian Hearne, said: 'It is a pity that this decision has taken this long as it has put MRM in a difficult position. The cost of the delay to our business and our suppliers' businesses is irrecoverable.'[124]

FROM MISSIONARIES TO MERCENARIES

From the 1960s onwards, large-scale mining proliferated in remote regions around the country, accelerating in the 1970s as the Poseidon boom gained momentum. In places like the Pilbara, mining took place as if Aborigines did not exist. If they were in the way, they were moved. At Mapoon on Cape York in 1963, an entire community was forcibly moved to make way for bauxite mining.[125] In the Northern Territory, the federal government suddenly realised the value of the resources and tried to take the land that had been granted to Aborigines. On 11 March 1963 the Menzies government set about resuming 220 square kilometres of the Arnhem Land Reserve to make way for mining by the Swiss-owned Nabalco on the Gove peninsula. The outcry among the Yolngu people culminated in the famous 'bark petition', which led to a parliamentary inquiry that recommended Aboriginal people be paid compensation for the impact of mining on their land.[126]

At the same time, BHP was eyeing the rich manganese deposits newly discovered on nearby Groote Eylandt, about 50 kilometres southeast of Gove. Groote Eylandt is home to the Anindilyakwa people and one of the world's richest manganese deposits. The stuff is not hard to find—just push back the topsoil and there is one of the essential minerals needed for steel-making. Since 1921 the evangelical Church Missionary Society (CMS), an arm of the Anglican church, had operated a mission on the island. Upon learning of the valuable mineral deposits the CMS took out prospecting rights over the entire island, which it claimed was for the benefit of the locals. Without any Aboriginal involvement, CMS and BHP struck a deal in 1963 that allowed the company to begin mining. Having been ruled by missionaries for almost half a century, the island was soon inundated by miners.[127] The mining lease negotiated in 1964 extended for twenty-one years, with the right of renewal for a further twenty-one. Soon after another lease was granted for mining on Gove that stretched for almost a century—forty-two years plus the right to negotiate a 42-year extension.

As these negotiations were being concluded, the Vietnam War triggered a boom in minerals prices, prompting a wave of investment that lasted until the global recession of the early 1980s. In the aftermath of this boom, a national study on mining in remote communities by David Cousins and John Nieuwenhuysen found that mining brought considerable upheaval and few tangible economic benefits to Indigenous Australians. The authors concluded that 'the economic and social impact of mining on remote Aboriginal communities has been great', but that employment results had been 'limited and unstable'.[128] They were surprised to find very little Indigenous employment in mining. Those jobs that were held by Indigenous employees tended to be unskilled. When economic downturn hit in the early 1980s, Aboriginal workers were the first to be laid off.

Some communities, however, succeeded in negotiating additional royalties for land access beyond what the companies paid in state royalties. On Groote Eylandt, the missionaries secured a royalty of 1.25 per cent, meaning that this share of the gross value of production was paid directly to the Aboriginal community. The Yolngu people on Gove, however, were unable to extract any additional payments from Nabalco. This was before the time of land rights legislation. But when the Fraser government passed the Aboriginal Land Rights Act (1976)—which had been developed by the Whitlam government—Aborigines in the territory gained de facto mineral rights over Aboriginal land by virtue of a clause that allowed them to veto mining projects. The law was followed by even better royalty deals. The Gagudju community near the Jabiru uranium mine negotiated a royalty of 1.75 per cent in 1978, while the 1982 Jabiluka Agreement committed 2 per cent, more than the amount that was to be paid to the Territory government.

The industry came to describe these deals as 'fuck-off agreements', and the results often proved detrimental. Bruce Harvey, who heads Aboriginal and community relations at Rio Tinto, describes the Ranger deal as the 'richest and worst deal ever'; it handed over a huge slice of royalties and in so doing created a 'generation of despair'. There is also evidence that the mining payments allowed the government to pull back on services, thereby shifting costs onto the communities. And generous agreements did not improve Aboriginal employment. By the early 1980s, Ranger employed just ten Aboriginal workers, or 2.6 per cent of the workforce, and Groote just twenty-one, or 4.4 per cent.[129]

Within the first decade of the Groote Eylandt deal, BHP-controlled GEMCO was producing 2 million tonnes of manganese a year from the tiny island. The company had unearthed one of the richest such ore bodies on earth, one which is still being mined. Production is now running at 4.25 million tonnes a year and in 2010 the operation earned BHP $2 billion. Professor Jon Altman of the Australian National University, who looked at the Groote experience in the early years, said the trust fund set up to manage mining royalties, which was controlled entirely by Indigenous leaders, took a conservative approach to spending the money during the first decade of mining. Initially the money was spent on community facilities, but over time this shifted towards 'ceremonies and dance festivals'.[130] Andrew McMillan, a Darwin-based journalist who has visited the island regularly over recent decades, writes that the people of the island arguably became the 'richest Aboriginal people in Australia', but that little was done by government, GEMCO or the self-governing community councils to ensure that mining improved the welfare of the people. Instead, the combination of royalties and welfare seemed to diminish the motivation of these islanders. While the mission system was highly paternalistic, McMillan writes, people had been compelled to work for their food, to tend the gardens or work for rations or go out hunting and fishing. But the flood of money proved to be debilitating. 'The motivational business of hunting or fishing or working for your tucker fell by the wayside.'[131]

GEMCO brought an influx of miners to the island. They soon came close to outnumbering the locals, and their social mores influenced local culture for the worse. The miners brought all of the prizes of Western civilisation: flash cars and boats, videos and alcohol. The effect of grog on the community led to the highest rates of imprisonment in Australia, twenty-five times the national average. Those who resisted Western temptations had to live with the effects of mining, with open-cut pits and ore-crushing operations located just a few hundred metres from one of the island's main settlements, Angurugu. Few of the locals on Groote found work at the mine.

In the Pilbara, development of the iron-ore industry from the 1950s occurred without any legislative support for the local Aboriginal people. When mining companies began staking claims in the early 1960s, the area had a white population of about 3,200 and an Indigenous population of about the same size, although they were not officially counted in the census until after the 1967 referendum. By the early 1980s, the iron-ore developments had increased the white population by a factor of fifteen to 47,000, leaving Aborigines marginalised.[132] Mining companies like Hamersley Iron, a joint venture between CRA Ltd (now Rio Tinto) and Kaiser Steel of the US, built mining towns to house the growing workforce that explicitly excluded Aborigines. BHP ran one of the biggest operations in the region, the Mt Newman Mine. With a workforce of more than 4,000 people, the mine should have presented an enormous opportunity for the local Aborigines. But by the early 1980s, after more than a decade of operation, MNM had engaged just fifteen Aborigines, or 0.4 per cent of the local Indigenous population—the lowest percentage identified by the Cousins and Nieuwenhuysen study. Not only were Aborigines not receiving any compensation payments for the intrusion of mining onto their traditional lands, they were almost entirely excluded from job opportunities.

REBEL MOMENTS

When several members of the Bougainville Revolutionary Army (BRA) shot and seriously wounded two Australian employees of CRA in 1989 and then killed a contractor, the experience proved to be a catalyst; the company soon transformed the way it did business with Indigenous communities in Australia. It could be argued that the rebels did as much to improve the welfare of Indigenous Australians as the Native Title Act. Peter Taylor, who worked at the mine for two years in the late 1980s, described Bougainville— home to one of the world's biggest copper and gold operations—as an idyllic place that at the time gave no hint of the trouble that was to arise. There were 'fantastic facilities', including an international school, for the 500-strong Western workforce. Unlike today's FIFO workforce, employees were encouraged to bring their wives and children to the island. And unlike some mining towns in Australia at the time, this township was an 'open town': it allowed locals to live alongside the miners and be part of the community.

But CRA omitted to pay compensation directly to the locals, and this was a significant factor in the rebellion. At the time the mine opened there was no provincial government, and so CRA negotiated with Port Moresby. Only a 'very small amount found its way back to the landowners', says Taylor, who now heads Bougainville Copper Ltd and has begun working towards restarting operations.

As Bougainville was unravelling, CRA began thinking hard about its operations in its own backyard. Previously, like most mining companies in Australia, it had paid little attention to Indigenous relations or welfare. One reason for this was that outside the Northern Territory, Aboriginal people had extremely weak negotiation rights. This all changed in 1992 with the *Mabo* case, when the High Court ruled that Aboriginal people could in some circumstances have common-law title to land extending back to the beginning of white settlement. Prime Minister Paul Keating then proposed the Native Title Act, which gave Aboriginal people the right to have their claims to native title heard by the National Native Title Tribunal and the Federal Court. The legislation was met with derision by business groups and the conservatives, especially mining and rural interests, who stoked fears of a black land grab. It became a shrill and ugly debate that threatened to divide the country. But CRA's experience in Bougainville had changed the company's philosophy. The chief executive of CRA, Leon Davis, broke ranks with the industry. In March 1995 he told the Securities Institute that his company saw 'major opportunities for growth in outback Australia, which will only be realised with the full co-operation of all interested parties'. He said the Native Title Act provided the basis for 'a series of CRA operations developed in active partnership with Aboriginal people'. In a second speech that April he said the company wanted to move away from a litigious framework and to establish 'innovative ways of sharing with and/or compensating Aborigines'.[133]

Like most other companies operating in Australia, Rio Tinto won't disclose the details of financial settlements with native-title groups, even though it is a party to global agreements that promote full transparency by resource companies. Some Aboriginal people complain that under agreements with Rio they end up with 'the richest trusts and the poorest people'.[134] However, figures on employment show that the company has moved ahead by leaps and bounds compared with the early days. Rio now employs more than 1,600 Aboriginal workers, making it the single biggest employer of Indigenous people in Australia. In its Argyle diamond mine about one in four workers is Indigenous, and in the Ranger uranium mine the figure is almost one in five. Aboriginal employment in Rio's iron-ore operations is about one in ten, a vast improvement on the situation a generation ago. In June 2011, the company finalised agreements with five communities in the Pilbara that should underpin substantial economic opportunities and development. Other mines big and small have followed suit. BHP has similar employment levels in the Pilbara, although the massive Olympic Dam operation, with its 3,000 BHP employees and contracters, has just 3 per cent Indigenous employment. Even this is a big improvement on the situation a decade ago, when BHP acquired Western Mining Corporation—reportedly there wasn't a single Indigenous employee. But it is unlikely that all or even the majority of these Indigenous workers come from nearby communities; many of them are flown in from major towns and cities.

Indigenous businesses are also providing opportunities, with a number of firms emerging as significant operations. Indigenous experts cite companies like Carey Consulting in the Western Australian goldfields and Bull Yanner at the Century Mine in Queensland as encouraging examples. In early 2011, the Eastern Guruma people in the Pilbara won a $160 million contract with a joint-venture partner. Rio Tinto's

Pilbara deals are aimed at building up local businesses as well as providing jobs.

Despite the good will shown by some companies, however, and the huge scale of developments in remote areas near Indigenous settlements, research shows that the benefits of mining still prove elusive for most Aboriginal people. The overall picture is one of mineral wealth making no discernible impact on the high rates of poverty experienced by Aborigines. Extensive research by Professor John Taylor, a human geographer at the Australian National University's Centre for Aboriginal Economic Policy Research, found that the socio-economic status of people living in the 'hinterland' of developments in Kakadu and the Pilbara is 'often indistinguishable' from other Aborigines in remote Australia. Taylor attributes this surprising result to such factors as the inability of people and organisations to cope with the impacts of large-scale mining and take advantage of them, and Indigenous peoples' 'ambivalent responses' to the assimilation required to work in the industry.

THE MORE THINGS CHANGE

Andrew 'Twiggy' Forrest has built a reputation for wanting to uplift the lives of impoverished Aboriginal people in Australia. He has used his wealth and influence to focus attention on Indigenous job generation through government-funded projects like the Australian Employment Covenant (AEC). And through his own company, the emerging iron-ore producer Fortescue Metals Group (FMG), Forrest has emphasised jobs for local Aborigines rather than paying the standard industry compensation packages in return for access to native-title land. Forrest is entirely disparaging of such compensation payments, describing them as 'mining welfare' that merely replaces government welfare.

Instead of cash, Forrest says, he is offering jobs to Aborigines who are long-term unemployed, although so far his Indigenous employment record, at about 8 per cent, is no better than that of his competitors. In pursuing this approach, FMG has shown a ruthless capacity to drive a very hard bargain in negotiations with native-title groups. The company has been accused of deliberately splitting communities and offering huge initial signature bonuses but very low levels of ongoing income.

One of Forrest's earliest deals involved access to the Chichester Ranges. A former native-title lawyer, Marcus Priest, has documented how FMG secured the signature of six Nyiyaparli elders, led by David Stock, for a compensation deal worth a minuscule 2.5 cents a tonne, plus signature fees worth almost $500,000. Stock had known Forrest since Forrest was a boy growing up on Minderoo station in the Pilbara; Stock taught him and his brothers to ride horses. Now that FMG is on its way to becoming 'the new force in iron ore', the company has shown a capacity to exploit such connections. In August 2005, Stock and five other Nyiyaparli native-title claimants signed an agreement to give FMG access to 40,000 square kilometres of their land. The agreement removed significant cultural heritage and environmental provisions that had been agreed between FMG and the Nyiyaparli in earlier negotiations, and it was signed without the elders having legal representation present. The broader claimant group disavowed the agreement two weeks later and Stock said he hadn't known what he was agreeing to. Stock was quoted in the media the day after signing: 'I didn't know what was going on. I feel like they made me sign—they kept calling me uncle.'[135] After going through a protracted legal claim to gain native title, the Nyiyaparli forfeited their rights as a result of this deal.

FMG's corporate affairs chief, Deidre Willmott, who joined the company after working on the staff of Western Australian Premier Colin Barnett, says that when other benefits are included the compensation paid is seven times the figure of 2.5 cents a tonne. Even if this is so, that still works out at 0.01 per cent of the production value, or one-fiftieth of the 0.5 per cent paid by Rio and BHP in comparable native-title deals at current prices. FMG's approach to these negotiations involved dividing the community by fanning frustration with the Indigenous negotiators, in this case the Pilbara Native Title Service (PNTS). Forrest claimed that Stock and his mates 'pushed their way into our office because of their absolute frustration of dealing with PNTS'.[136]

In subsequent negotiations, FMG has used a similar approach. The lawyer Ronald Bower, principal of the firm Corser & Corser, wrote of earlier negotiations with the Eastern Guruma people: 'Our own experience of negotiations with FMG on behalf of the Eastern Guruma native-title holders was very slow and frustrating, with FMG being resolutely committed to what our clients and we considered at the time to be extremely mean-spirited terms …'[137]

However, Bower is now representing a group known as the Wirlu-Murra Yindjibarndi Aboriginal Corporation, which has broken away from the main Yindjibarndi claimant group to deal with FMG. The Yindjibarndi people, who number 1,200, succeeded in 2005 in having their claim to native title recognised by the Federal Court at Roebourne. The claimant group was originally represented by the Yindjibarndi Aboriginal Corporation (YAC). FMG wants access to Yindjibarndi's determination area and an extended claim in order to produce 60 million tonnes a year in iron ore, worth around $10 billion, more than doubling its current output. In return, FMG has offered a cash settlement of $4 million a year (with no indexation) plus $6.5 million in unspecified jobs and training benefits. The cash offer works out at 0.04 per cent, less than a tenth of the offer of 0.5 per cent made by Rio to other Pilbara groups, according to YAC chief executive Michael Woodley.

The approach taken by FMG in these negotiations has many of the hallmarks of the earlier Nyiyaparli agreement— a big upfront settlement fee followed by a modest dollar amount rather than a percentage share of the value. And there are claims that FMG has been working to split the community. Since early 2010, FMG has had an anthropologist based at Roebourne Art Group, which has become the headquarters for the rival Wirlu-Murra group.

Ronald Bower (who is indirectly paid by FMG, given the company funds the Wirlu-Murra) says of the proposed arrangements: '[The Wirlu-Murra] wish to enter into an agreement with FMG which, although it promises to provide a comparatively very poor level of financial benefits to them in terms of compensation or payments in the nature of royalties, would provide their families with opportunities for training, employment and business development.' As a sweetener, FMG is offering a signature fee of $500,000 and has been paying Wirlu-Murra people sitting fees of $500 each to attend meetings and vote in favour of the deal. FMG's Deirdre Willmott says such fees are standard in the industry.

While Yindjibarndi leaders claim that FMG has deliberately divided their community, FMG says it was already divided. Some members, the company claims, were unhappy with the lack of consultation by Michael Woodley, a charismatic and highly articulate individual who appears to have made a few enemies. Woodley, now thirty-eight, left school at sixth grade and was educated by his grandfather in Yindjibarndi law. He is steeped in traditional culture, but he understands the modern world and released a video of a March 2011 native-title meeting that revealed some of FMG's tactics. The company bussed in about 160 people to a staged meeting, where each attendee signed their name in order to be paid a sitting fee. FMG succeeded in having the video removed from the original website (which was hosted by a company in New York), but the Yindjibarndi then put it up on YouTube.

Wirlu-Murra leader Allery Sandy says the dispute in the community began with a 'family feud', as all of the

Yindjibarndi are in fact related. Sandy declines to say how much FMG has paid her group, and when asked if she knows about more substantial offers involving a percentage share of production, she says, 'None of our people know what a percentage is.' The comment affirms the real fear raised by the YAC's legal adviser, barrister George Irving— that the breakaway group supports giving FMG or any third party far-reaching access to their land on terms that they do not understand.

Elder Bigali Hanlon says the breakaway Wirlu-Murra group has been influenced by the upfront money, which won't go very far in the Pilbara. Instead, she says, long-term compensation should be used to educate children in the community and start businesses. Hanlon is talking from experience, for her three children are university educated and now work as professionals. One of them is Jody Broun, the former head of the New South Wales Department of Aboriginal Affairs, co-chair of the National Congress and a successful artist who has won the Telstra Aboriginal and Torres Strait Islander Arts Award. Hanlon also believes there should be compensation for the untold damage mining would do to traditional lands. The community is being asked to sacrifice 'prime sacred sites' in an ancient river bed. This will mean the loss of sacred stones used in initiation ceremonies. Hanlon says other mining companies like Rio have paid fair compensation and have 'not been so ruthless'.[138]

Tough tactics have also been used against the Western Australian government, which has tried to resolve the dispute between the YAC and the Wirlu-Murra by bringing the two groups together. When the Department of Aboriginal Affairs hosted a mediation session in February 2011, Ronald Bower took legal action against a departmental official, Brian Wilkinson, alleging that Wilkinson had described him as a 'servant of FMG'. Wilkinson denies the allegation. As a result of the legal action, the department has had to curtail its mediating role.[139]

While the state government has at least tried to intervene, the federal government sees no role for itself in raising the standard of native-title negotiations. The emphasis so far has been on scrutinising recipients of native-title deals, rather than the mining companies. This was the thrust of a 2010 paper on native-title reform released by the attorney-general, Robert McClelland, and the Indigenous affairs minister, Jenny Macklin, although McClelland has, when questioned, acknowledged that 'more information' should be made available by mining companies about the money that is paid during negotiations and in final settlements.[140] And Martin Ferguson, despite being a former trade unionist, doesn't see a need for any minimum standards or parameters in these negotiations. 'They have to work that out. The main thing from my point of view is the accountability of foundations and how they spend the money. They employ well-paid lawyers and advisers to work out the appropriate packages. Not for us to be dictating to them how to do their job,' he says. The result of this laissez-faire approach is that companies can use opaque arrangements to gain control of native-title land. Governments demand more and more accountability of Australia's most disadvantaged people, but seem uninterested in demanding the same ethical standards of those with massive wealth and bargaining power.

Most of the big resource companies operating in Australia, and the federal government, have signed up to a global anti-corruption regime known as EITI—the Extractive Industries Transparency Initiative—which enshrines the principle of 'publish what you pay'. While EITI is aimed at helping developing countries, where such projects have been a major source of corruption, first-world nations like Norway are among the thirty-five resource-rich countries to have signed on to the agreement. The regime involves full disclosure of company payments and government receipts for resource projects. Australia pays lip-service to these principles overseas, but none of our global commitments is being implemented at home. Without stronger ethical requirements, Australia's resources stampede could trample the Indigenous communities on whose land it depends.

How to Ride the Resources Rollercoaster

'We are in grave danger if mining runs into a wall, having shelled out the rest of the economy on the back of the high dollar … We would be setting ourselves up for a mega-bust.'—Foreign Minister Kevin Rudd, February 2011[141]

Without stronger policies, Australia is in danger of falling into the resource trap. Managing this boom is not rocket science. There is much we can learn from ourselves, from our own history, and from the extensive list of good practice examples around the world.

SAVE WHILE THE SUN SHINES

An important first step is to establish a savings fund for windfall revenue in a broad range of major foreign currencies including US dollars, euro and yen. We should do this immediately, while the dollar is high. There is no need to wait until the federal government has paid off all of its debt, as the accumulated savings would offset that debt. Whether it is called a stabilisation fund or a sovereign wealth fund doesn't really matter; what is important is that we make a start and set some ground rules for how it would operate. What begins as a short-term stabilisation fund could grow into a wealth fund as the boom unfolds. We already have a working model in the Future Fund.

Supporting such a fund has become like saying you are in favour of lower interest rates (something the fund would help to achieve). A lot of people now support the principle, but the hard part is defining how the fund should work and ensuring it doesn't get raided by a future government ahead of an election. Creating the fund now is important because the government's financial position will improve dramatically should the mining boom roll on at its current pace.

SPEND THE AVERAGE

Although no system is perfect, overseas models suggest some steps we might take. Norway's approach is a good start because it makes sure that oil revenue isn't frittered away in the annual budget. But the government has to work out the net present value of every oil field in the country to arrive at the sustainable rate of expenditure. For Australia, with hundreds of resource projects underway at any one time and many more on the way, this would be a monumental task, and one that is not really necessary. Norway's spending limit of 4 per cent is also too low for Australia, as our growing population demands higher investment in public works projects. The Chilean approach, meanwhile, has a serious flaw: politicians can easily avoid reaching the tax revenue benchmark at which they must start saving by giving revenue away through populist tax cuts. The model that provides the best example for Australia is a lot closer to home—just north of the Torres Strait, in fact.

After earlier attempts, the PNG government adopted a set of rules to manage its resource revenue based on 'spending the average'. It is a fantastically simple but effective piece of policy; the Treasury officials who thought of it deserve Orders of Australia. It doesn't rely on trying to forecast revenue, which is well-nigh impossible. Instead, it simply uses history as a guide. Here's how it works. Take the average share of GDP for all federal resource revenue for a long period, say twenty years. This becomes the spending limit for the present generation. Any additional revenue is automatically saved in what might be called the 'Australian Resources Future Fund'. When the mining boom abates and its contribution to revenue drops, the government can draw on the fund by continuing to spend the average rate of resource revenue to boost growth. If the boom lasts for decades as predicted, the average will steadily rise, allowing us to spend more in a sustainable way, without having to rely on heroic forecasts. Spending the average is like an automatic stabiliser for the economy. Some economists think this model is too conservative for a developing country like PNG, and the government recently changed the rules so that all of the above-average revenue is available for infrastructure investment. Nonetheless, the principle is a good one for Australia and the PNG experience underlines the need for robust rules.

POLLIE-PROOF THE SAVINGS

Given the weaknesses of our political system, there is always the risk that governments will raid the fund. We've seen too much of this in Australia and it's time to end it. The solution involves an act of parliament that restricts spending of resource revenue to the twenty-year average; any plans to spend more than this must gain the unanimous approval of three institutions: the federal parliament, the secretary to the Treasury and the board of the Reserve Bank. If we got really serious, we could follow Alaska's example and propose constitutional change to enshrine these principles for all time, although it's unlikely we'd muster sufficient political will to achieve this.

STATES SHOULD BE SAVERS, TOO

State governments should also be accumulating future funds with a share of their mining royalties, and they too should adopt the principle of spending the average. There are some alarming examples of how the states are blowing their resource revenue. Despite an enormous lift in royalties, the 2011 Western Australian budget revealed that state debt will reach $20 billion in two years, five times the level of 2008. One of the problems is that the rules governing Commonwealth grants provide no incentive for states to run more rational and effective tax regimes for mining, or to accumulate savings. Any increase in saving or revenue means Commonwealth grants are cut. As David Murray says, 'There is a flaw in that structure that doesn't encourage the states to set aside money'. This disincentive needs to be removed—in return for a commitment from the states to save a share of their royalty income.

A UNIFORM RESOURCE RENT TAX

Uniform and effective taxation of mining profits is needed to accumulate savings during the boom years, and to achieve more balanced development. It clearly makes no sense to have the petroleum industry paying a 40 per cent resource rent tax while coal and iron-ore projects pay 22.5 per cent and all other mining pays nothing. Without uniform taxation, we will see more of the poorly thought-out investment that we have been getting of late. The last proposal from Treasury was more complex than it needed to be. We have a workable model in the Petroleum Resource Rent Tax, which has not discouraged more than $130 billion of investment in LNG exports in recent years. As with the PRRT, the new uniform tax would tax profits at a rate of 40 per cent after the companies have benefited from a generous write-off of exploration costs. Extending the PRRT across the entire resources industry would not kill the golden goose—it would allow the goose to live a lot longer and lay many more eggs.

MAKE MINING SAFE AND FAMILY FRIENDLY

The FIFO model being employed by most operations is bad for regional communities and the development of Australia's interior, and it is bad for families stuck at home without a mum or dad for an extended spell. The drive-in, drive-out model is also becoming a danger to the community, given the emerging evidence of road accidents caused by exhausted miners returning home at the end of several back-to-back shifts. The government should look at taxation arrangements that encourage companies to locate workers permanently in mining towns. Safety standards for the industry should extend to cover how miners return home at the end of long shifts. It is not acceptable for companies to leave workers to manage fatigue unassisted. Employers should be responsible for making sure their workers can commute safely. Solutions could include providing buses or taxis to take workers to and from work, much in the way city workers who finish late at night are guaranteed a safe ride home in a taxi.

REVEAL AUSTRALIA'S TRUE WORTH

When resource companies dig up minerals, extract oil and gas from the ground or harvest timber, Australia's GDP records these activities as a contribution to output, even though our endowments of natural resources have in fact declined. Despite being a country increasingly dependent on natural resources for its export income, Australia has no aggregate data available on the extent of these resources. ABS data only records the flow of income in our national accounts; the impact on our stock of the assets underpinning these activities is left a mystery. The same is true of Treasury's account of the nation's wealth. The omission gives the false impression that such resources will last indefinitely.

Australia should create a national balance sheet that shows all natural resource endowments, together with existing data on financial and physical assets. This idea is supported by Ken Henry, who has said it is 'a challenging concept for a country with a relatively high endowment of natural resources', but so far nothing has been done.[142] Creating such a balance sheet is a fairly straightforward exercise, given that the information is readily produced by Geoscience Australia. But this balance sheet should be forward-looking as well, showing the impact of future contracts and expansion plans. We should also recommence the collection and publication of data on the foreign ownership of Australia's mining operations.

REGULATE JOINTLY

Just as the Northern Territory was found grossly inadequate in its oversight of a relatively minor, shallow-water oil field in 2009, so too will other smaller administrations prove unable to monitor and regulate massive resource projects. Joint regulation by the federal and state governments is clearly needed, as advocated by Chloe Munro. We shouldn't have to rely on resource companies to fund environmental studies. This is the responsibility of government.

Joint regulation should also involve developing, monitoring and enforcing national safety standards. We need such standards to limit the ultra-fine dust emitted by mines, the chemicals used to frack coal seams, and the water consumption of mining and coal-seam gas operations. The federal government has ignored these harmful side-effects for too long. As damage often flows across borders, it is essential for both levels of government to regulate together.

INDIGENOUS PEOPLE SHOULD BENEFIT, TOO

It will be more than a crying shame if this boom doesn't help to raise up the Aboriginal people who are mired in poverty while living on top of industrial mining complexes. It is not good enough for our politicians to say that Aboriginal people should be left to fend for themselves in negotiations over land rights with multinational corporations. Some minimum standards are needed, just as we have minimum wages for low-paid workers. Companies should pay a percentage rather than a fixed dollar amount, with funding linked to education, training and employment opportunities, as well as setting some money aside for future generations. An industry standard is needed and the Rio Tinto model would be a good place to start. Companies should be required to fully disclose their agreements with Aboriginal communities and the amount of money and other benefits given to individuals and communities during negotiations, just as they should be fully transparent in all their payments.

PUBLISH WHAT YOU PAY

Many major mining companies have declared their support for the Extractive Industries Transparency Initiative but do not practise it in Australia. These companies include BG, BHP Billiton, Chevron, Gold Fields, Mitsubishi Minerals,

Newmont, OZ Minerals, Santos, Shell, Total, Woodside and Xstrata. The federal government has also pledged support for EITI, but it has not required that all resources companies comply with the regime. Given its support for EITI, there's no reason why this should not happen.

A SOBERING REMINDER

For a perspective on where this boom might be taking us, it's worth looking to lessons from previous mining booms. The current investment stampede is starting to resemble Australia's second mining boom, which relied heavily on London investors and bankers and ended in misery in the late 1890s. So intense was this rush that from 1894, on average one Western Australian company was floated in London every day for a period of two years. In one month, April 1896, eighty-one were floated. As the RBA deputy governor, Ric Battellino, has observed, most of this money was never repaid in dividends, 'an indication of the risks that can be involved in mining investment'.

Geoffrey Blainey documents the excesses of this period in *The Rush That Never Ended*. The narrative is one of frenzied gambling rather than investment, with shady operators fleecing investors through fabrication. London stockbrokers travelled to the remote town of Coolgardie, about 50 kilometres east of Kalgoorlie, in search of new opportunities. For a few years the town was flowing with French champagne and British companies bought land in the main street, believing the place was destined to become a major city.[143] One mine the size of a grave raised 700,000 pounds when floated, but the investors later found out that what gold there was had been stolen. Blainey also describes the mining boom that swept the southwestern Tasmanian town of Queenstown in the late 1890s, where 'prospectors, financiers, swindlers, investors, clairvoyants and carpet-bag speculators' congregated to exploit the riches found underneath nearby Mt Lyell. 'The Tasmanian press called it Copperopolis, the copper city. In Queenstown, in December 1897, it was said that you could not meet a businessman who was not dabbling in mining shares.'[144]

Mt Lyell proved to be a vast source of copper and silver; so great were its metallic riches that it is believed to have had an effect on Abel Tasman's compass when he sailed along the west coast of the island in November 1642. In one lucky strike, a quarter of a million ounces of silver came from a slope barely the size of a suburban dining room. It proved to be so rich that it attracted one of the four original members of the Broken Hill syndicate, Irish immigrant Bowes Kelly. The parallels with the present boom are uncanny. Before the First World War, most of the ore from Mt Lyell and Broken Hill was shipped to the world's emerging industrial power at the time, Germany, just as most of our ore today is being shipped to China.[145]

Loaded with the proceeds of the Broken Hill development, Kelly and his fellow board members proved themselves to be enthusiastic developers and gamblers. 'They gambled for profit and they gambled for pleasure, playing two-up with gold sovereigns and much laughter at the close of the weekly board meeting,' writes Blainey. Kelly continued to gamble on mining ventures in Tasmania, most of which never made money. At the end of his life he lived in a dilapidated but ornate Italian-style mansion next to the old Government House in Glenferrie Road, Melbourne, owning a fraction of his former wealth.[146]

When Blainey returned to Mt Lyell in the 1950s, he was hopeful that more ore would be found. He urged the government to 'assist' the company in such endeavours, declaring that such a mine could still be paying dividends at its centenary. He feared that without such efforts, Queenstown could become a ghost town:

> Even if all the available copper is blasted out until only a deep crater remains at West Lyell, the field may not perish. Eminent geologists are confident that an intensive drilling campaign will find more copper ...
>
> If the price of copper were to fall and to remain low for years, if no new ore body were found, and if governments refused to assist the company, it is within the bounds of reason that Queenstown might become a sleeping village haunted with old fossikers and pensioners before the century closes.
>
> But there are happier portents of the future—the brighter prospects in the open cut, the systematic attempt to find new ore bodies, Australia's rapidly expanding export market for copper, and the chance that new methods of mining or smelting may make even poorer ore payable.

And so the field might be mining one hundred years after the bold prophecy made by a Melbourne shareholder during the share slump in 1898 that, when London is in ruins, the Mt Lyell furnaces even then will be in blast and paying dividends to the descendents of all who are wise enough to bequeath their present holding to their children.[147]

After a century of mining at Mt Lyell, an entire valley that was once a lush rainforest is now completely denuded, pockmarked with craters filled with toxic tailings. The devastation covers an area of about 50 square kilometres, the result of 'pyritic' smelting operations and timber harvesting to feed the growth of the town early last century. Mt Lyell has long ceased paying dividends, though the Indian company Sterlite (also on Norway's ethical blacklist) bought the assets when the previous owner went broke and runs a small underground operation. The town is a shadow of the glory days described by Blainey; the population is about 2,000 and a three-bedroom home can now be bought in Queenstown for just $50,000. There is nothing to show for the vast riches extracted over the past 108 years.

The state of Tasmania was paid enormous royalties for almost a century, but its leaders never thought to save for when mineral prices collapsed, or for when the ore ran out. Mt Lyell and Tasmania are harbingers for what Australia as a whole might look one hundred or even fifty years from now.

ENDNOTES

1. 351 mining complexes … Figures taken from australianminesat-las.gov.au, using 'Quick Search: Mines and Mineral Deposits, Operating Mines, All Commodities'. Last accessed 1 July 2011; One billion tonnes … *Australian Mineral Statistics 2011*, December Quarter 2010, Canberra: ABARES, p. 11.
2. Estimate of $334 billion … Laurie, K. and McDonald, J. 'A Perspective on Trends in Australian Government Spending', *Economic Roundup*, Canberra: The Treasury, Summer 2008, pp. 27–49.
3. Corbett comments … 'Resolving the mining tax agreement is urgent, says Corbett', *The Australian*, 3 February 2011.
4. 15 per cent estimate … Stevens, G. 'The Challenge of Prosperity', address to the Committee for the Economic Development of Australia, Melbourne, 29 November 2010. Referring to a graph of Australia's terms of trade over the past 130 years, Stevens said: 'This means that about 12 to 15 per cent of GDP in additional income is available to this country's producers and/or consumers, each year, compared with what would have occurred under the average or trend set of relative prices over the preceding 100 years (all other things equal). That will continue each year, while the terms of trade remain at this level.' The New Zealand extrapolation is the author's.
5. Stevens and TVs … *Ibid.*
6. Grenville's stampede … Grenville, S. 'What did I change my mind about in 2010? The mining boom and the deficit', *The Interpreter* (Lowy Institute blog), 4 January 2011, www.lowyinterpreter.org.

7.   Martin Ferguson ... interview with the author, 22 February 2011.

8.   Investment and consumption ... *Government Finance Statistics,* Catalogue Number 5512.0, Canberra: ABS, various years. Historical data reproduced by the ABS using accrual accounting.

9.   Geoscience Australia estimates ... *Australia's Identified Mineral Resources 2009,* Canberra: Geoscience Australia.

10. Gas estimates ... *Australian Energy Resource Assessment,* Canberra: Geoscience Australia and ABARE, 2010.

11. Tripling of gas production ... Syed, A., Melanie, J., Thorpe, S. and Penney, K. *Australian Energy Projections to 2029–30: ABARE research report 10.02,* Canberra: ABARE, 2010.

12. Australia to become second biggest ... The Office of the Chief Economist of the International Energy Agency. *Are We Entering a Golden Age of Gas? Special Report, World Energy Outlook 2011,* Paris: IEA, 2011, p. 69.

13. 116 million tonnes increase ... Syed, A. et al. *Australian Energy Projections to 2029–30,* p. 45.

14. Australia ranked twelfth ... Central Intelligence Agency. 'Natural Gas – Proved Reserves', *The World Factbook,* https://www.cia.gov/library/publications/the-world-factbook/rankorder/2179rank.html, 2010.

15. Bob Gregory ... Gregory, B. 'Some Implications of the Growth of the Mineral Sector', *Australian Journal of Agricultural Economics,* Volume 20, Number 2, August 1976, p. 71. Gregory is one of the author's PhD supervisors.

16. White elephants and Angola ... 'Spread the Wealth', *The Economist,* 10 February 2011.

17. Stevens study ... Stevens, P. and Dietsche, E. 'Resource Curse: An analysis of causes, experiences and possible ways forward', *Energy Policy,* Issue 36, Volume 1, January 2008, p. 58.

18. Zambia ... Ferguson, J. *Expectations of Modernity,* Berkeley: University of California Press, 2009, pp. 9–14.

19. Infant mortality figures ... *Ibid.,* p. 12.

20. Nauru ... examples of excess from former advisers who worked in the country.

21. RBA paper ... Clifton, K. 'The effects of large increases in capital inflow for Australia', June 2010. Obtained by the author under FOI law.

22. Peak Downs Highway ... author interview with Noel Lang, retired policeman and chair of the local Road Accident Action Group, June 2011.

23. Andrew Fraser ... interview with the author, March 2011.

24. Ricardo ... Ricardo, D. *On the Principles of Political Economy and Trade*, Cambridge: Cambridge University Press, 1982, pp. 132–5.

25. Shepparton speech ... Stevens, G. 'Monetary Policy and the Regions', address to Foodbowl Unlimited Forum Business Luncheon, 20 September 2011.

26. Professor Paul Ehrlich ... interviewed on *Late Night Live*, ABC Radio, 19 November 2009.

27. Swan briefing ... Treasury Executive Minute, 30 July 2009.

28. Pilbara skills shortages and Chris Adams ... Stutchbury, M. 'The raw edge of the boom', *The Weekend Australian*, 13 October 2010.

29. RAN shortages ... 'Creatures of the deep', *The Australian*, 26 June 2008.

30. Long-run price trends ... O'Connor, J. and Orsmond, D. 'The Recent Rise in Commodity Prices: A Long-run Perspective', *Bulletin*, Sydney: RBA, April 2007.

31. 64 million vacant apartments ... Brown, A. 'China's Ghost Cities', *Dateline*. SBS Television, 2011.

32. Investment pipeline ... New, R., Ball, A., Copeland, A. and commodity analysts. *Minerals and Energy Major Development Projects – April 2011 listing*, Canberra: ABARES, May 2011; 70 per cent figure ... '5625.0 – Private New Capital Expenditure and Expected Expenditure', ABS, March quarter 2011, http://www.abs.gov.au/ausstats/abs@.nsf/mf/5625.0.

33. Katherine Woodthorpe ... interview with the author, June 2011.

34. David Gruen … 'The resources boom and structural change in the Australian economy', speech to the Committee for Economic Development of Australia (CEDA), 24 February 2011. Available online at treasury.gov.au.

35. Tourism graph … Tourism Research Australia. *State of the Industry 2010*, Canberra: Department of Resources, Energy and Tourism, p. 9.

36. Arrivals and departures … *Overseas Arrivals and Departures, March 2011*, Catalogue Number 3401.01, ABS, released 10 May 2011.

37. Unemployment rates in Cairns and the Gold Coast … Labour Market Research and Analysis Branch, Labour Market Strategy Group. *Small Area Labour Markets Australia*, Canberra: Department of Education, Employment and Workplace Relations, December quarter 2010.

38. Jason Anderson … interview with the author, December 2010.

39. Mitch Hooke comments … 'Claims of 2-speed peril untrue', *The Australian*, 12 January 2011.

40. Chris Evans … *7.30*. ABC Television, 13 April 2011.

41. Chinese students drop 50 per cent and $5 billion figure … Sainsbury, M. 'Decline in China numbers persist', *The Australian*, Higher Education Supplement, 15 December 2010.

42. NIEIR research … NIEIR. 'The Economic Costs of the Current Expenditure of Mining Expansion: Policies for Cost Minimisation', unpublished research paper for the Steel Institute Victoria, 2011, p. 4.

43. China and India growth figures … Australian Government. 'Statement 4: Opportunities and Challenges of an Economy in Transition', *Commonwealth Budget Paper No. 1*, 2011–12, pp. 22–28.

44. Steel intensity figures … Battellino, R. 'Economic Developments', address to CEDA, 18 November 2010.

45. Iron ore expansion … 'BHP unveils ambitious $48bn Port

Hedland iron ore expansion plans', *The Australian*, 19 April 2011; Rio figures … author interview with Sam Walsh, chief executive of Rio Tinto Iron Ore, June 2011.

46. Brian Fisher … interview with the author, April 2010, and email communication, March 2011.

47. Glenn Stevens … 'The Resources Boom', speech to Victoria University Public Conference on the Resources Boom: Understanding National and Regional Implications, 23 February 2011. Available online at rba.gov.au.

48. IEA forecasts … 'IEA sees gas oversupply until 2020', Reuters, 10 November 2010; OCE of the IEA, *Are We Entering a Golden Age of Gas?*

49. Wood MacKenzie report … Garvey, P. 'Cheap US gas a looming threat', *The Australian*, 13 June 2011.

50. Worthwhile for government to invest … Garnaut, R. and Clunies Ross, A. *The Taxation of Mineral Rents*, Oxford: Clarendon Press, 1983, p. 66.

51. African investment … 'Strategic resources: a richer seam', *Financial Times*, 20 May 2010; and 'China seeks big stake in Nigerian oil', *Financial Times*, 28 September 2009.

52. Sixty pubs in Broken Hill … Broken Hill City Council. *Broken Hill: The Accessible Outback*, 17th ed., p.7.

53. Stevens' speech on volatility … Stevens, G. 'The Road to Prosperity', address to the 2009 Economic and Social Outlook Conference Dinner, Melbourne, 5 November 2009.

54. When asked about it later … Remarks made by Stevens following his initial address, 5 November 2009.

55. Stephen Grenville … interview with the author, April 2011.

56. Response from Swan via email, August 2010 and February 2011.

57. Joe Hockey … interview with the author, February 2011.

58. Tony Abbott … interview with the author, June 2011.

59. Phillip Blond … interview with the author, June 2011.

60. Andrew Leigh … interview with the author, April 2011; 'The case

against a sovereign wealth fund', *Australian Financial Review*, 28 June 2011.

61. David Murray ... interview with the author, June 2011.

62. Garnaut paper ... Garnaut, R. 'Breaking the Australian Great Complacency of the Early Twenty-First Century', address to the 2005 Economic and Social Outlook Conference. Garnaut's comments about the response to this speech were made in an interview with the author, May 2008.

63. Treasury minute to Costello ... 'Meeting with the IMG Regarding the 2007 Article IV Visit', 18 June 2007. Obtained by the author under the FOI law.

64. Treasury paper ... McCissack, A., Chang, J., Ewing, R. and Rahman, J. *2008–01: Structural Effects of a Sustained Rise in the Terms of Trade*, Treasury Working Paper, July 2008.

65. Several decades ... For one example see Henry, K., 'The Shape of Things to Come: Long-run forces affecting the Australian economy in coming decades', address to the Queensland University of Technology Business Leaders' Forum, 22 October 2009. Available online at treasury.gov.au.

66. Chile's economic collapse ... Auty, R. *The Resource Curse Thesis*, London: Routledge, 1993, p. 116.

67. *Ibid.*

68. Andrés Velasco ... *Informe Anual Fondos Soberanos 2009*, Santiago, Chile: Ministry of Finance.

69. Peter Costello ... Costello, P. 'Bad luck enters with a boom in parallel reality that is Wayne's world', *The Sydney Morning Herald*, 27 April 2011.

70. David Murray ... interview with the author, June 2011.

71. NIEIR paper ... NIEIR, 'The economic costs of the current expenditure of mining expansion', p. 4.

72. Albanese warning ... 'Rio chief uses Rudd case as a warning', *The Australian*, 10 July 2010.

73. Ken Henry's boyhood ... Henry, K. 'Economics, Economists and

Policy', presentation to the Masters of Economics Thirty-Year Reunion Symposium, Australian National University, 21 September 2001. Available online at treasury.gov.au.

74. Kevin Rudd ... interview with the author, 15 February 2011.

75. Kevin Rudd ... *Ibid.*

76. Freyburg warning ... *The World Today*. ABC Radio, 8 September 2010.

77. Geoff Walsh ... interview with the author.

78. Newspoll ... 'Federal voting intention and leaders' ratings', 21 June 2010. This poll, taken over the period 18–20 June, four days before Rudd was deposed, showed the ALP leading the Coalition on a two-party-preferred basis by 52 to 48 per cent.

79. Barnett comment ... *Q&A*. ABC Television, 6 November 2010.

80. Garnaut on Bougainville ... interview with the author, March 2011.

81. Howard comment on PRRT ... 'Treasurer's Economic Note', 9 May 2010. Available online at treasurer.gov.au. Last accessed 1 July 2011.

82. Downer comment on PRRT ... Buckingham, D. 'Behind the resource rent tax hysteria', *Business Spectator*, 10 May 2010.

83. Craig Emerson... 'Iron-clad case for revamped resources tax', *The Weekend Australian*, 3 July 2010.

84. Tony Maher ... interview with the author, April 2011.

85. Cedric Marshall ... interview with the author, June 2011.

86. John Rolfe study ... Rolfe, J., Lawrence, R., Gregg, D., Morrish, F. and Ivanova, G. 'Minerals and Energy Resources Sector Economic Impact Study', Brisbane: The Eidos Institute for the Queensland Resources Council, 2010, p. 22; and John Rolfe ... interview with the author, June 2011.

87. Fiona Rossiter ... interview with the author, March 2011.

88. RBA study ... Connolly, E. and Orsmond, D. 'The Level and Distribution of Mining Sector Revenue', *Bulletin*, Sydney: RBA, January 2009, pp. 7–12.

89. *Ibid.*, p. 9. Note that this paper omits a figure for taxes paid by the miners because the publication was not released at the time the paper was written. The relevant figure was taken from the same publication used by the RBA for the earlier estimates.

90. Minerals Council of Australia ... return on capital and equity figures for thirty years to 2009, unpublished document.

91. Rio foreign ownership estimates ... Rio Tinto. 'Submission to the Senate Standing Committee on Economics', Canberra, 2009, p. 17.

92. ABS survey ... *Foreign Ownership and Control of the Mining Industry*, Catalogue Number 5317.0, ABS.

93. Naomi Edwards ...'Foreign ownership of Australian mining profits: Now are we selling the farm?' Briefing paper prepared for the Australian Greens, released 30 June 2011.

94. Several decades ... The first of many uses of this term was made in Henry, R. *The Shape of Things to Come: Long-Run Forces Affecting the Australian Economy in Coming Decades*, Canberra: The Treasury, 22 October 2009, p. 13.

95. Chloe Munro ... interview with the author, April 2011.

96. Mining position paper ... *Mining Position Statement: The Mining and Water Challenge*, Canberra: National Water Commission, May 2010.

97. Dr Wayne Smith ... interview with the author, March 2011.

98. Shenhua spends $200 million ... 'Chinese mine giant snaps up 43 NSW farms', *The Australian*, 27 June 2011.

99. Fiona Newell ... interview with the author, March 2011.

100. Paul Keating ... 'Gas company accused of bullying', AAP, 16 March 2011.

101. Burke's advice ... 'Proposed Decisions – Australia Pacific LNG Project', DSEWPC, 2 February 2011, p. 2. Obtained by the author under FOI law.

102. Geoscience Australia advice ... *Ibid.*, pp. 1 and 8.

103. Salinity ... *Ibid.*, p. 8.

104. Narran Lakes ... *Ibid.*, p. 9; 'Supporting Advice – Wetlands Section', DSEWPC, p. 2. Obtained by the author under FOI law.

105. IPCC forecasts ... *IPCC Fourth Assessment Report: Climate Change 2007 (AR4)*, Chapter 11.

106. CSG position paper ... *Position Statement: The Coal-Seam Gas and Water Challenge*, Canberra: National Water Commission, 3 December 2010.

107. Garrett letter ... Attachment B, letter to Penny Wong, 12 July 2010, included as an appendix in a submission from the Department of Sustainability and Environment to Tony Burke, 27 September 2010. Obtained by the author under FOI law.

108. Hamilton paper ... Hamilton, C. and Downie, C. 'University Capture: Australian Universities and the Fossil Fuel Industries', The Australia Institute, Discussion Paper Number 95, June 2007.

109. Chris Moran ... interview with the author, June 2011.

110. Hillier comments ... Carney, M. 'The Gas Rush', *Four Corners*. ABC Television, 21 February 2011.

111. Graham Clapham and Ruth Armstrong ... Lloyd, G. 'Fertile ground for coal-seam test case', *The Australian*, 21 May 2011.

112. Lock the Gate ... Lloyd, G. 'Gates shut in the gas lands', *The Weekend Australian*, Inquirer, 28–29 May 2011.

113. Tailings storage ... Draft *Environmental Impact Statement (EIS)*, BHP, Volume 1, Chapter 5, 2010, p. 128.

114. Ranger regulations ... *Ranger Project Environmental Requirements*, 14 November 1999, Appendix A to the Schedule to the authority issued under section 41 of the *Commonwealth Atomic Energy Act 1953*.

115. Affect groundwater 6 kilometres below ... Draft *EIS*, BHP, Volume 1, Chapter 12, p. 361.

116. Olympic dam pit and tailings runoff ... *Ibid.*, pp. 361–4, and Table 12.3, 2010.

117. Rudd response ... Response to Questions on Notice by Senator Scott Ludlam, 6 December 2010.

118. Don Henry ... interview with the author, March 2010.

119. BHP response ... email to the author from a BHP spokeswoman, June 2011.

120. Supplementary EIS ... *Olympic Dam Project: Supplementary Environmental Impact Statement*, BHP, 13 May 2011.

121. Ammonium nitrate ... *Ibid.*, p. 42.

122. Bigali Hanlon ... interview with the author, April 2011.

123. 60 per cent figure ... *Indigenous Relations – Strategic Framework*, Canberra: Minerals Council of Australia, December 2004.

124. Brian Hearne ... 'Xstrata Zinc Australia welcomes today's draft decision by Minister Garrett on MRM', Press Release, 22 January 2009, http://www.xstrata.com/media/news/2009/01/22/0956CET/.

125. Mapoon community moved ... Altman, J. 'Indigenous communities, miners and the state in Australia', *Power, Culture, Economy*, J. Altman and D. Martin (eds), *CAEPR Research Monograph 30*, Canberra: ANU E-Press, 2009, p. 24. Altman is one of the author's PhD supervisors.

126. Bark petition and Menzies' resumption of land ... Altman, J. *Aborigines and Mining Royalties in the Northern Territory*, Canberra: Australian Institute of Aboriginal Studies, 1983, p. 17.

127. CMS and BHP ... *Ibid.*, pp. 11–12.

128. Cousins and Nieuwenhuysen ... Cousins, D. and Nieuwenhuysen, J. *Aboriginals and the Mining Industry*, Sydney: Allen & Unwin, 1984, pp. 167–8.

129. Aboriginal employment numbers ... *Ibid.*, p. 2.

130. Spending GEMCO royalties ... Altman, *Aborigines and Mining Royalties*, p. 16.

131. McMillan comments ... McMillan, A. *An Intruder's Guide to East Arnhem Land*, Nightcliff: Niblock Publishing, 2007, p. 195.

132. Pilbara population ... Cousins and Nieuwenhuysen, *Aboriginals and the Mining Industry*, p. 120.

133. Leon Davis speeches ... Harvey, B. 'Rio Tinto Agreement Mak-

ing in Australia in a Context of Globalisation', *Honour Among Nations?*, M. Langton, M. Teehan and L. Palmer (eds), Melbourne: MUP, 2004, pp. 239–40.

134. Richest trusts … Scambary, *My Country, Mine Country*, p. 165.

135. FMG's Nyiyaparli deal … Priest, M. 'Brand New Day', *Australian Financial Review Magazine*, July 2006, pp. 41–9.

136. Pushed their way in … *Ibid.*

137. Ron Bower … letter to Ronald Bower from Slater & Gordon, January 2011.

138. Bigali Hanlon … interview with the author, April 2011.

139. Bower allegations and denial … author interview with Ronald Bower and email response from departmental spokesman, May 2011.

140. Response from McClelland via email, May 2011

141. Kevin Rudd … interview with the author, 15 February 2011.

142. Henry comments on balance sheet … Henry, K. 'Fiscal Policy: More Than Just a National Budget', address to the Whitlam Institute, Sydney, 30 November 2009.

143. Blainey on Coolgardie … Blainey, G. *The Rush That Never Ended*, 3rd ed., Melbourne: MUP, 1978, pp. 188–90.

144. Blainey on Queenstown … Blainey, G. *The Peaks of Lyell*, 2nd ed., Melbourne: MUP, 1959, p. 84.

145. Most ore shipped to Germany … *Ibid.*, p. 280; Blainey, *The Rush That Never Ended*, p. 297.

146. Bowes Kelly … Blainey, *The Peaks of Lyell*, pp. 266–7.

147. Mt Lyell … *Ibid.*, pp. 287 and 290.

ACKNOWLEDGEMENTS

Thanks to Chris Feik of Black Inc. for strongly supporting this work and for excellent ideas and comments on the draft manuscript, and to my agent, Lyn Tranter, for introducing me to him. Denise O'Dea of Black Inc. and Simon Draper provided helpful comments on the manuscript. Peter Fray deserves special thanks for having suggested the title. Several people have provided ideas that made their way into this book: Jon Altman, Glen Boreham, Peter Colley, Simon Hawkins, Bob Gregory, Ross Garnaut, Brad Law, Warwick McKibbin and Simon Nish. Paul Roberts and Luisa Ryan of the Australian Bureau of Statistics produced a special data series on public-sector spending. Thanks to the editors of the *Australian*, Chris Mitchell, Clive Mathieson and Nick Cater, who have given me the scope to initiate coverage of some of the issues developed in more detail in this book. Thanks to people from various organisations who went out of their way to provide material, access to their busy bosses or other fonts of information: Amanda Buckley, Claire Muntinga, David Noonan, David Luff, Ben Mitchell, Fiona Scott, Dean Souter and Kerrina Watson. I am especially grateful to the Freedom of Information officers in the Department of Sustainability and Environment, the RBA and the Treasury. And to KP, who read drafts and provided comments, and Cheeky Chops, for the 'wow' factor.

ABOUT THE AUTHOR

Paul Cleary is a senior writer with the *Australian* and a PhD research scholar in public policy at the Australian National University. During the decade he was based in the Canberra press gallery, his reporting on tax reform in the late 1990s used Freedom of Information law to expose the impact of the proposed GST on low-income earners. This led to a Senate inquiry and to beneficial changes for these taxpayers. He was awarded a Chevening Fellowship by the UK Foreign Office to study at SOAS at the University of London and subsequently worked for the government of the newly independent East Timor, where he was involved in negotiations over disputed oil resources in the Timor Sea and in setting up the country's petroleum fund. He is the author of two previous books, *Shakedown: Australia's Grab for Timor Oil* and *The Men Who Came Out of the Ground*. He speaks Vietnamese and Tetum. To contact the author about this book, email orestruck@gmail.com.